Flexible Project Management for Product Development

Your illustrated guide to making project management work in tumultuous development

4th Edition

By: José Campos

Published by:

Rapidinnovation Press

www.Rapidinnovation.com

Flexible Project Management for Product Development (4th Edition)

By: José Campos

Published by:
Rapidinnovation Press
Rapidinnovation, LLC
25000 SW Big Fir Rd.
West Linn, Oregon 97068
United States
www.rapidinnovation.com

ISBN-13: 978-0-9817595-3-1
ISBN-10: 098175953X

Published in the United States

CIP data available from the publisher

Get your interactive PDF version FREE!
See the last pages of this guidebook for details

V-FLXPM/111017/6.2.3/ACS/PRN/SW

Introduction

Flexible Project Management

Critical Definitions and Concepts

Advanced Communication

Team Building and Teamwork

Four Phases of Flexible Project

The Typical Development Program

Introduction

Phase 1: Define the Project

Phase 2: Plan the Project

Introduction

Phase 4: Close Out the Project

Critical Tools

Deployment of a New Tool

Problem-Solving Process

Karnaugh Maps: 2x2 Matrixes

Appendix B: Glossary

Appendix C: FMEA Process Map from SEMATECH, Inc.

Appendix D: Templates

For comments, questions or additional copies please e-mail the author: flexpm@rapidinnovation.com.

Legend

Navigate through this Guidebook

To ease navigation, we have provided a set of icons, which provide orientation and clarity. They are consistent throughout the guidebook, always carrying the same meaning.

Symbols to Indicate Facts to Know

Important Suggestion

Used with each important suggestion, note, or tip, derived from experience in the field during actual projects.

Essential Quote

A quick way to read very important information. We suggest you read all essential quotes.

Flexibility

Indicates the flexibility of a particular tool. We encourage you to adapt the tool to your specific situation and not be governed by rigid, prescriptive frameworks.

Definition

A defined term. See also our Glossary at the back of the guidebook.

Time Frame

To help you plan the likely duration of a meeting, work session, or other event.

You Are Here

One more way to get oriented or to check that you have take the proper steps up to this point.

Symbols for Links to Helpful Info

Note for Print Version

Interactive features are available in the electronic PDF version. If you have a printed version, contact the authors to learn how to upgrade your guidebook.

Template (Printed Version)

You will find copies of the templates at the back of your guidebook.

Template (Electronic Version)

Double click this icon to open a corresponding template (enabled for use in Adobe Reader and Acrobat) that you can fill with your own information and share with your peers and colleagues. When you close the template, you will return to this guidebook.

Each time you open a template from your guidebook, it will be fresh. But you can save versions that you have completed, partially filled in or left blank for future access. Saved versions remain enabled for revisions and collaboration.

These enabled templates also provide you with Comment & Mark-up Tools. Just as you might with a hard copy, you and your team can highlight text, add notes and make custom annotations.

See Also (Electronic Version)

Clicking this icon will jump you to a place in the guidebook with more information relevant to the topic at hand.

Back Button (Electronic Version)

After following a link, you can easily click to return to your original place in the guidebook.

Introduction

Approach

Why We Wrote this Guidebook

Project management for product development has unique characteristics and challenges from traditional project management. In addition, we feel that there are methods that are not contained in traditional project management but are very relevant for product development. Thus, we set about creating this guidebook to provide Program Managers, Project Managers and Engineering Leads as well as those leading development projects a more relevant resource, written by people who understand both project management and product development.

The mission of this guidebook is to enable you to apply project management methods to product development, where the environment is complex due to constant change that happens in developing new products.

We want to help you become more productive, increase performance, and deliver results. Results are products that are on time, on budget and on profit. Results can be measured in terms of market share and competitive advantage, but always leading to profitability.

A How-To Manual

Having been on the receiving end of guidebooks, handbooks, "primers" and training during our lives in the corporate world, we realized that often the information was not sufficient for us to actually use it in our situation, or the information was formatted poorly such that it made it impossible for us to apply it, which resulted in much frustration, not to mention loss of productivity. Worse yet, the documentation we received from training sessions was simply a hard copy of the PowerPoint slides shown during the session, which were of little help to us when it came time to apply the learning.

We made a commitment that our guidebook would contain everything a professional would need to apply every concept in it. We have strived to provide the tools, methods and encouragement for you to successfully understand and apply the voice of your customers.

How to Use this Guidebook

We developed this guide for individuals and teams who are in product development (Marketing and R&D) and have specific assignments to deliver results. The entire focus of the guidebook is to lead you — step-by-step — through each and every concept, even if you have never performed the task before.

The sequence of this guidebook is based on a chronological approach — the outline and format follows a logical "starting point" and move towards a "end point." This means that the higher the page number, the later in the process.

This is not a book about concepts. We have limited concepts to the minimum necessary to create context or perspective. The focus of this guidebook is on the "how-to."

The Process Maps

The most important element of this guide is the process maps — the step-by-step notes for starting and completing key tasks. These process maps lead to the various tools and templates.

There is a hierarchy of process maps:

> **LEVEL 1: THE MASTER PROCESS** This is the highest view of the process. The Master Processes, also known as Mental Models, represent big pictures, and they are rarely actionable due to their strategic nature. They are simply here to give you the overall perspective of a large endeavor or system.
>
> **LEVEL 2: MAIN PROCESS** These are the major phases contained in the guidebook. They provide a phased (stage) approach and are indeed actionable. Main Processes as generally good for planning purposes, also to discuss with your team as you prepare to start a particular stage.

LEVEL 3: DETAILED PROCESS These are the processes you will need to start and complete tasks. They not only contain the steps, but we have also included critical information in the form of tips, suggestions to help you locate the progress and other important information, please see the Legend at the start of this guidebook for further details on the icons used to help you navigate.

The Template and Tools

Templates are forms that contain information that you and your team enter in order to collaborate and create archival information for later use. Tools are matrices, spreadsheets, tables and other instruments.

It is important to note that for every tool or template there is at least one example and in some cases several. This is intended to provide you with a general idea of the information that should be entered. It is not intended to limit what you and your team feel needs to be done.

Unique Capability for the Interactive PDF Version

Not only does every template in this book have a full example, but also it is fillable. You and your team can enter information about your project and share it. You will notice a link in the form of an icon and when you click on it, you will be able to open a blank template that you and your team can use. Once you have completed entering the information, the template becomes its own PDF independent of the guidebook. You can save it using your own name or title, which you can also share with the rest of your team around the world. You can continue entering information until you are satisfied.

Essential

"The most important element of this guidebook is the process maps — the step-by-step notes for starting and completing key tasks."

Flexibility

We understand that every development project is different — by definition. Consequently, our intent is not to limit your flexibility. All the process maps, templates and tools can be modified to better meet your objectives and style. We suggest that the first time you apply the these process maps, templates and tools that you try to adhere to our format. As you become familiar with the intent, you can modify them until you have something that is unique to your organization and also meets your requirements to the fullest. Don't be afraid to experiment and modify.

Mission of this Guidebook

The mission is to enable you and your team to apply Project Management methods and tools to deliver new products on-time and on-profit.

Project Management is a powerful methodology to ensure that you meet your time-to-maket targets and, more importantly, that you meet your time-to-profit goals through the achievement of the desired gross margin for your new product.

Objectives of this Guidebook

The overall objective is to create the capacity for improved productivity! And the results of using this guidebook will include these specific objectives:

- You will be familiar with overall strategies needed for flexible project management to develop new products on time.
- You will be able to apply the material presented in this guidebook in real-world situations.

Your Input

No book is perfect; this one is no exception. If you find difficulties in using the tools, concepts and processes, do not hesitate to contact the authors: flexpm@rapidinnovation.com.

Introduction

About the Author

José Campos, Principal Consultant and Founder, Rapidinnovation

José collaborates with B2B technology companies around the world on the methods, tools and cultural shifts to rapidly and consistently deliver superior value to customers.

José has spent his career working to define and deliver innovative products to market. For 30 years he has passionately focused on discovering new methods and approaches to discover and deliver superior solutions to the customer in the tumultuous world of high-tech

He has written several acclaimed books: *Voice of the Customer for Product Development* — a methodology to discover, prioritize, and use customer requirements to create clearly differentiated products — faster than the competition. Also, *Risk Management for Product Development* — the strategies and methods to reduce the impact of uncertainty on your investment. Jose was coauthor of Project Management Toolbox in collaboration with Dr. Dragan Milosevic.

His focus on high-tech and his technical background allow him to provide meaningful and practical coaching to technology professionals. He has traveled the world help R&D and Marketing teams improve their ability to create innovative products that customers love.

Related Guidebooks for Product Development

In addition to *Risk Management and FMEA for Product Development*, our library includes several related guidebooks in the Product Development series:

- The Voice of the Customer for Product Development
  Your illustrated guide to obtaining, prioritizing and using customer requirements and creating winning products.
- Risk Management and FMEA for Product Development
  Your illustrated guide to reducing time-to-market through risk management and FMEA.
- Actionable Metrics for Product Development
  Your illustrated guide to developing and using metrics to improve product margins and reduce time-to-market.
- Debriefs and Postmortems for Product Development
  Your illustrated guide to improving performance through lessons learned.

Training, Consulting and Facilitation

Our highly experienced authors are also available to provide direct consulting to your organization. They can be team facilitators to help you implement the contents of these guidebooks. They can provide training on the topic plus many other areas of product development. Contact us for more information.

For comments, questions or additional copies please email us: jose@rapidinnovation.com.
Visit us at www.rapidinnovation.com.

Get Your Interactive PDF Version FREE!

See the last pages of this guidebook for details on how to acquire your free copy of the interactive PDF version.

Flexible Project Management

Critical Definitions and Concepts

NEW TOPIC

Important: Read Me First

Your Practical Tool Box

We wrote this guidebook with you in mind. It has considerable amount of content, tools, methods and processes all based on our experience working with development teams around the world. We have included all the information you might need as you lead product development programs.

Because we have included so much information, this guidebook may at times appear intimidating. Please, fear not! We added all this information because it is difficult to predict which parts you will apply to your situation. We leave it up to you to select your tools from this tool box. This is the essence of being flexible. And the good news is that you can pick the right tools for each project and set aside the rest for later use.

Thinks of this guidebook as your resource center, where you select those elements that will help you and your team deliver better products in a shorter period of time.

How to Approach Product Development

Whether a seasoned project manager or someone with little or no experience in project management, perhaps even new to product development, we encourage you to first read the **CRITICAL DEFINITIONS AND CONCEPTS** section. You can start by applying those methods, processes or tools that might be of immediate help to you and your team. Then start implementing the concepts using the Four Phases of Flexible Projects, perhaps gradually until you gain enough confidence to develop your own tool box and improve your current process. As you become familiar with this content of this guidebook, you can add more methods, processes and tools to continually improve your current approach.

Your Objective

Never lose sight that your role is to lead a development team to deliver new products that achieve the time-to-profit goals! Tools, methods and processes are simply enablers to help you with your leadership challenge.

Critical Definitions and Concepts

Complexity and Product Development

> Essential
> "Start using the word *complexity* to define product development instead of words like *chaos* or *confusion*."

Introduction

At one time or another we have all referred to product development as chaotic, confusing or even crazy. This is due to the many variables that can interfere with product development, each of which can disrupt the schedule, the budget and even the definition of your new product.

First of all, let's start using the word *complexity* to define product development instead of words like *chaos* or *confusion*. It's a more appropriate definition. Plus, it enables us to define product development projects in terms of their complexity ranking.

Second, you need to recognize that managing product development projects, especially in the high-tech industry, is inherently more complex than in other situations. Project management has its roots in the construction industry, where there is more predictability. But the number of variables and the frequency of change is more acute in product development, especially in the high-tech industry. Our point is that product development is difficult to manage because these variables interfere with traditional project management methods that assume a stable environment, and as you know product development is anything but stable.

Possible Sources of Complexity in Product Development

There are many sources of complexity. Case in point, consider the way budgets are managed and also the way decisions percolate across the company. The strategic intent or strategic initiatives may be defined by the executives of the company, yet the funding is allocated to functional groups such as marketing, engineering, etc. Now other priorities beyond those set by the executives set in. There is also the proverbial political landscape that aggravates the problem. Also factor in poor communication, financial woes, broken processes, etc. Confusion on top of chaos — no wonder things are complex.

Below is a short list of more possible sources of complexity in product development. Please read it and relate your current situation to the list. Naturally, you can add to and adapt this list.

Sources of Complexity	Examples
1. More than one organization within the same company involved in the development.	• Engineering, marketing and manufacturing involved in the same project. • People from two different divisions co-developing a new platform.
2. Members of your team are involved in more than one project.	• Engineers or marketing people working on several projects at the same time. • Engineering is also debugging the last "new" product.
3. Team members leave or arrive during the project.	• Layoffs, turnover, contractors, other projects, or consultants. • Team members are assigned to other projects.
4. More than two time zones involved.	• Team members in Seattle, Chicago and the New York. • Team members in India, Singapore and the U. S.
5. More than two national cultures involved.	• Team members in India, Singapore and the U. S.
6. More than two companies involved.	• Strategic partnership by two companies to develop a new product. • Your company has hired two engineering consulting firms to help develop the new product.

Critical Definitions and Concepts

Sources of Complexity	Examples
7. Not all members have the same technology tools.	• Some have high-speed internet, while others are on dial-up. • The simulation software are from several different vendors.
8. The scope changes, or is not clearly defined due to changes.	• Anticipating for a competitor to make its move, marketing decides to add a new feature. • A recent downturn in sales forces your company to cut your budget.
9. The technology used as part of the project changes.	• The new product and the technology for the new product being developed at the same time. • The entire team is using a brand new technology.
10. The market place is in turmoil.	• New technical standards. • Your competitors are very active.
11. A new project, which was not planned, is approved for immediate implementation.	• A major new customer requires a modified product. • A competitor pre-announces a new product.
12. Cost reduction forces change and complexity.	• Business is down; the company forces an immediate reduction in costs.
13. The financial model has high risk elements.	• You are not certain that your program will be profitable. • Too many uncontrolled variables in your financial model.
14. Your supply chain is weak.	• Your key suppliers are in financial trouble. • Your key suppliers are not certain that they can meet your timeline.
15. Your new product is going into a brand new market.	• Confusion about features and distribution channels, coupled with a lack of awareness of your company and your new product.
16. Your new product is going into your current market, but it is a breakthrough product.	• Confusion about the human interface; customers don't understand all the benefits. • Internally, there is a push to make it look like the old product.
17. Executives have placed tremendous pressure to get the new products to market, and the pace is breakneck.	• Simply put, getting the product to market in half the time you believe it should take. • The decision is made to debut at a product show which occurs two months ahead of your scheduled due date.
18. The program to get the new product to market has considerable complexity. It has new technology and new embedded software.	• Considerable interaction between each of the components of the new product — integration will be a nightmare. • The technology is new to your development team.

Critical Definitions and Concepts

Possible Sources of Complexity in Product Development

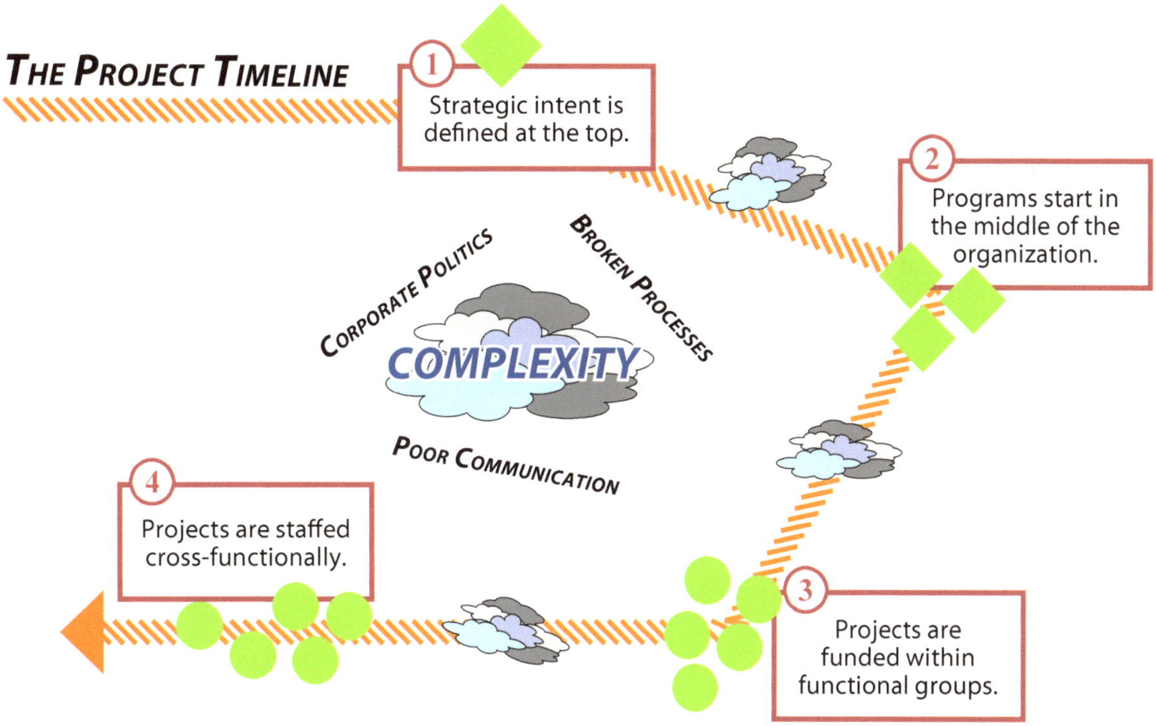

This diagram shows additional sources of complexity in product development. Some of these sources are self inflicted. For example, a broken process that interferes with the flow of information. Another example is the lack of clear communication. Corporate politics also add to the complexity.

The Importance of Flexible Project Management

In order to manage the complexity, remain flexible! Flexibility gives you the leverage to manage the variables that can interfere with product development. Planning ahead, being proactive and using tools and methods that accommodate complexity will allow you to be more flexible.

In the high-tech industry, the new product is simultaneously being developed with the new technology to be used in the product. It's difficult to schedule innovation. The ability to *flex* is critical.

Also, the corporate world can change; for example, your company could be acquired and all budgets are frozen. In which case, you need to be proactive by understanding how the system of funding and decisions works in your company and eager to change in order to meet your development objectives.

To lead development projects more effectively, as much as possible you need to plan ahead and be proactive. For example, define funding at the start, if not prior to the start of your project. Do your best to define the decision making process, document it and distribute it to all stakeholders. Maintain two-way communication throughout — make it a priority as your project depends on clarity of communication.

Moreover, the tools and methods you use in product development must be adaptable to account for its complexity.

Critical Definitions and Concepts

Every Project is Different

Every project is different; therefore, the tools, methods and overall approach will change with every project. The judgment of the project manager plays a critical role in selecting the overall strategy. Projects vary in complexity; for example, a project to remodel a conference room is relatively simple. On the other hand, a project to define a new product is significantly more complex. In addition, other variables influence the overall strategy the project manager must select; for example, the amount of technical innovation. A project using a mature technology is not as complex as one where the technology is being refined at the same time you are using it in your project. Other variables may include the number of people involved in your project, the geographic distribution of the team members, the size of the budget and the amount of risk associated with the project. In short, one size does not fit all. You must assess the selection of tools, method and overall approach in view of the variables that will impact your project. Avoid the cookie-cutter approach, where no flexibility exists between one project and the next one.

The Importance of Being Flexible

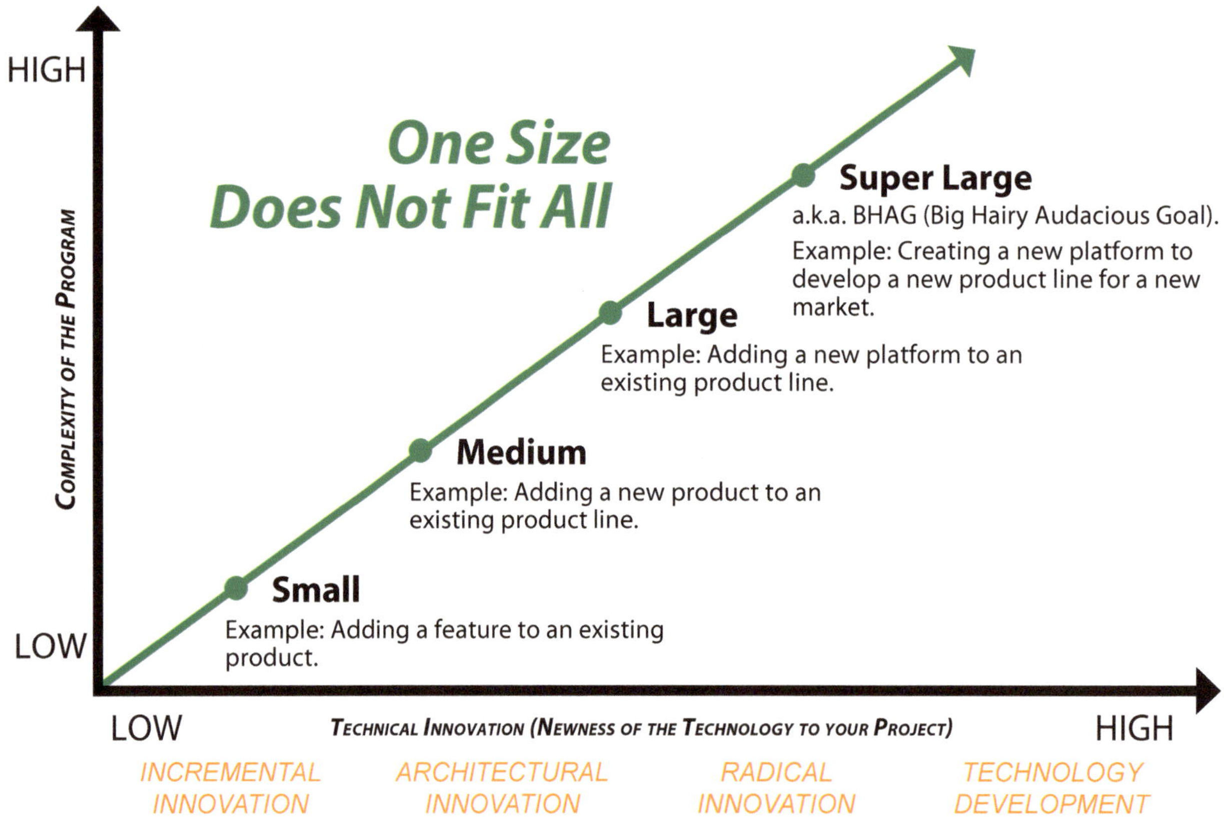

This diagram is relational, not functional. It symbolizes the scalability of projects in product development. For example, developing a new platform is more complex than adding a simple feature to an existing product; consequently you must scale your efforts and style to fit the size and complexity of your current product. There is no cookie cutter approach in product development due to their complexity. You must remain flexible and adapt to the current challenges.

Project Complexity Diamonds

Starting Point

Every organization faces different types of complexity, unique due to its culture, products, markets and business goals. Therefore, we encourage you to develop your own scale of complexity and avoid the one-size-fits-all definition. It is critical that your criteria for complexity be fully documented *and* supported by all stakeholders. Otherwise, its value is substantially reduced. We encourage you to take your time and allow for proper dialog in defining complexity for your organization — it is not a trivial endeavor. The benefits in clarity of communication and your ability to prepare for the complexity of any project before it starts will outweigh the initial effort.

The simplest way to define your complexity is to develop a set of criteria for each item. We suggest that each have no more than three elements to start. Later, you can add more substance. Approach it as if you are creating a dictionary or glossary. The scales are qualitative and not quantitative; therefore, we recommend that you and your team agree on the significance of each level in your scale. This is critical to the success of your complexity diamond. If there is no agreement in your team, there will likely be little support for the importance and meaning of your diamond.

The following table defines the axes common to complexity diamonds. You can develop your own set of variables as needed, but we recommend you follow the original set created by Dr. Aaron Shenhar as they have been tested in many projects; e.g., Novelty, Newness of Technology, Project Complexity, and Pace.

Term	Explanation
1. *-X axis* Novelty of the new product to your customers.	The focus is the product's newness to your customers and market. This axis should impact the product definition process; it would tend to be more difficult and take more time than defining a current product. This includes the newness of the technology as well as the product. For example, you may include a new technology in an existing product, which would constitute a higher complexity.
2. *+X axis* Newness of the technology to your development team or technology newness.	It is the uncertainty associated with the new technology used in the new product. A high level of uncertainty at the start of the product definition indicates considerable complexity for technology management, testing, development, etc. Newness may be applied to the technology itself, the newness to your company or organization or a technology new to your industry. For example, while a technology may not be new, it may be new to your design team, which may imply complexity. The opposite is a mature technology, which is stable and one you have used in several previous products.
3. *+Y axis* Project or program complexity.	This axis refers to the complexity of the project itself — for example: • The number of organizations involved in the development project. • The number of companies involved. • The number of vendors, consultants and suppliers involved. • The number of time zones or countries involved. • The size of the project in terms of dollars or the number of people involved.
4. *-Y axis* Pace or urgency.	How much time is available, versus how much time should it take. In a competitive environment, we do not always have the luxury of time. For example, missing an important industry trade show or conference may have severe financial impact on your company; e.g., your competitors showed new products and you did not. Getting to market first imposes a fast-paced approach, which naturally adds complexity to your project.

Critical Definitions and Concepts

Shenhar's Model of Project Complexity

Dr. Aaron Shenhar pioneered the idea of characterizing the challenge on a new project by describing its complexity. This approach clearly showed that all projects are different and consequently the approach to manage them should be different. This is particularly poignant in product development.

Shenhar's Three-Axes Model of Project Complexity

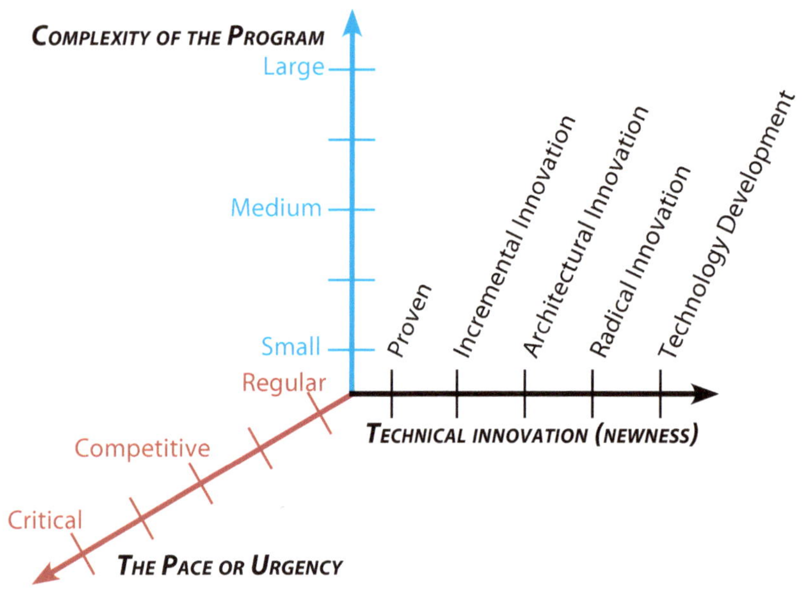

This approach uses three variables, each on its own axis.

COMPLEXITY: The number of variables in a project and their interconnection. For example, the number of people on the project, distance among the team members or number of layers of approvals.

PACE: The urgency and criticality of the time objectives in a project. For example, a fast-time-to-market product, restoring a mission critical network or an extreme competitive situation.

TECHNICAL INNOVATION: The familiarity that the development team has with the technology and the stability of the technology. For example, a new operating system used for the first time by the development team or a technology that is not fully developed and is still in the research laboratory.

Shenhar's Four-Axes Model of Project Complexity

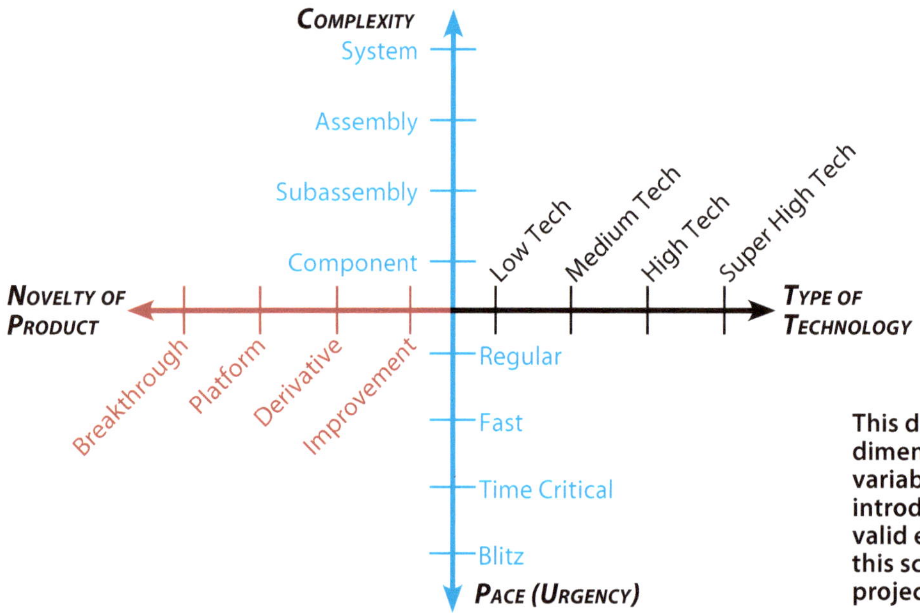

This diagram shows four dimensions of complexity. The variables on each axis were first introduced by Shenhar and are valid even today. You can use this scale and axis to map your projects.

For comments, questions or additional copies please e-mail the author: flexpm@rapidinnovation.com.

Critical Definitions and Concepts

Important

We wish to thank Dr. Shenhar for inspiring our application of his Diamon approach. Dr. Shenhar wrote *Reinventing Project management: The Diamon Approach to Successful Growth and Innovation* (Harvard Business press).

Real World Model of Project Complexity

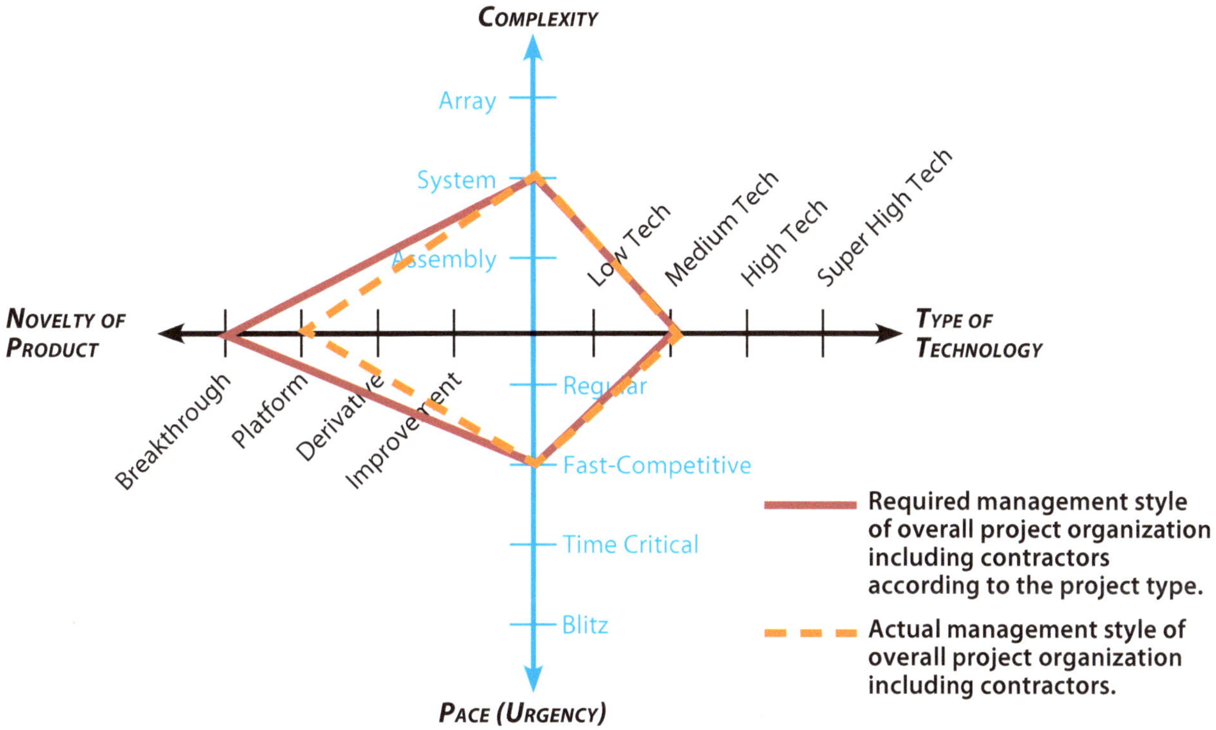

This diagram was developed for a real-world, very large project. The gap between the solid red line and the dashed orange line represents high risk. It indicates that the current capabilities of the organization (dashed line) don't match the demands of the project (solid line); consequently, it will overwhelm the organization and may result in project failure. Every time there is a gap between the demands of the project and the capabilities of the organization it requires corrective action before the project starts.

Critical Definitions and Concepts

Why Use the Complexity Diamond

- It is a graphical tool that enhances communication. In addition, it provides a more objective view of the challenge that confronts you; thus allowing for better planning and risk mitigation.
- It is a good team-building tool, as it creates alignment within the team.
- It will give you the design strategy that you will use. For example, in some projects you might use the relay approach, while others require the concurrent approach.
- It influences the engineering design strategy.

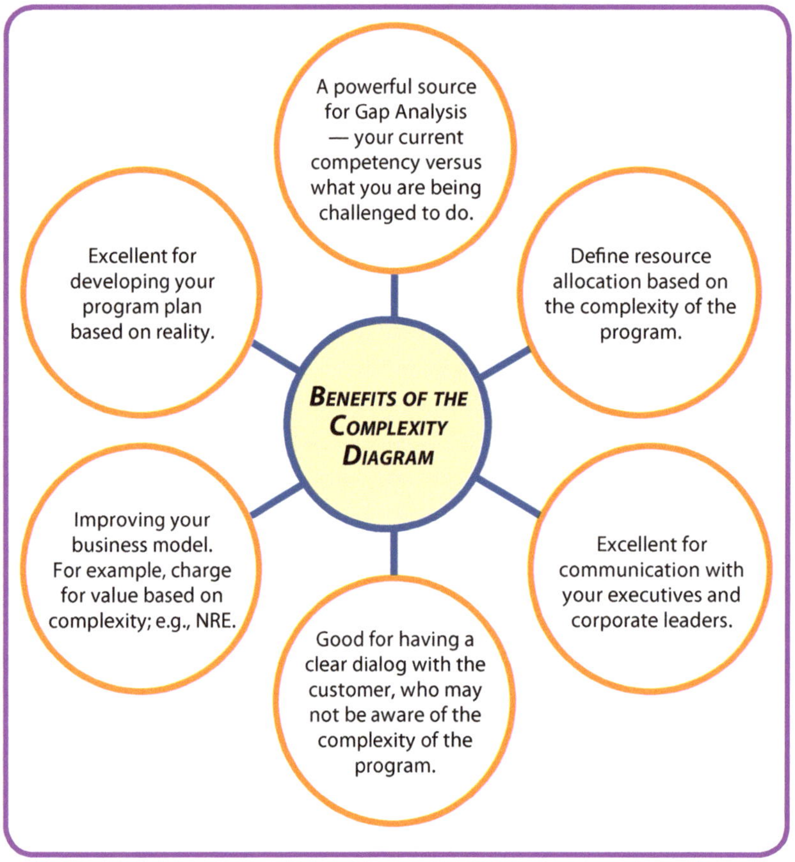

Critical Definitions and Concepts

Examples: Complexity Diamond

This example can be used to develop your own complexity diamond. We encourage teamwork — your individual perspective of the complexity of a project may be biased, but your team can bring clarity.

Scale

You can develop your own scales with the type of definition that makes sense for your business objectives and your project. We suggest that you use simple, linear scales from 1 to 10.

Other variables

Instead of the four variables, you may change them to others that better address your business objectives of your business. For example:

- **COST:** A high cost project may be more complex than a low cost one. There may be cases where cost is critical and you include it into your complexity diagram.
- **STRATEGIC IMPORTANCE:** The complexity of a very strategic project may be more than that of a non-strategic project.
- **CLARITY OF REQUIREMENTS:** A project with very unclear requirements will be considerable more complex than one where the requirements a clear, well-documented and reflect the true needs of your customer.

Shenhar's Model of Project Complexity

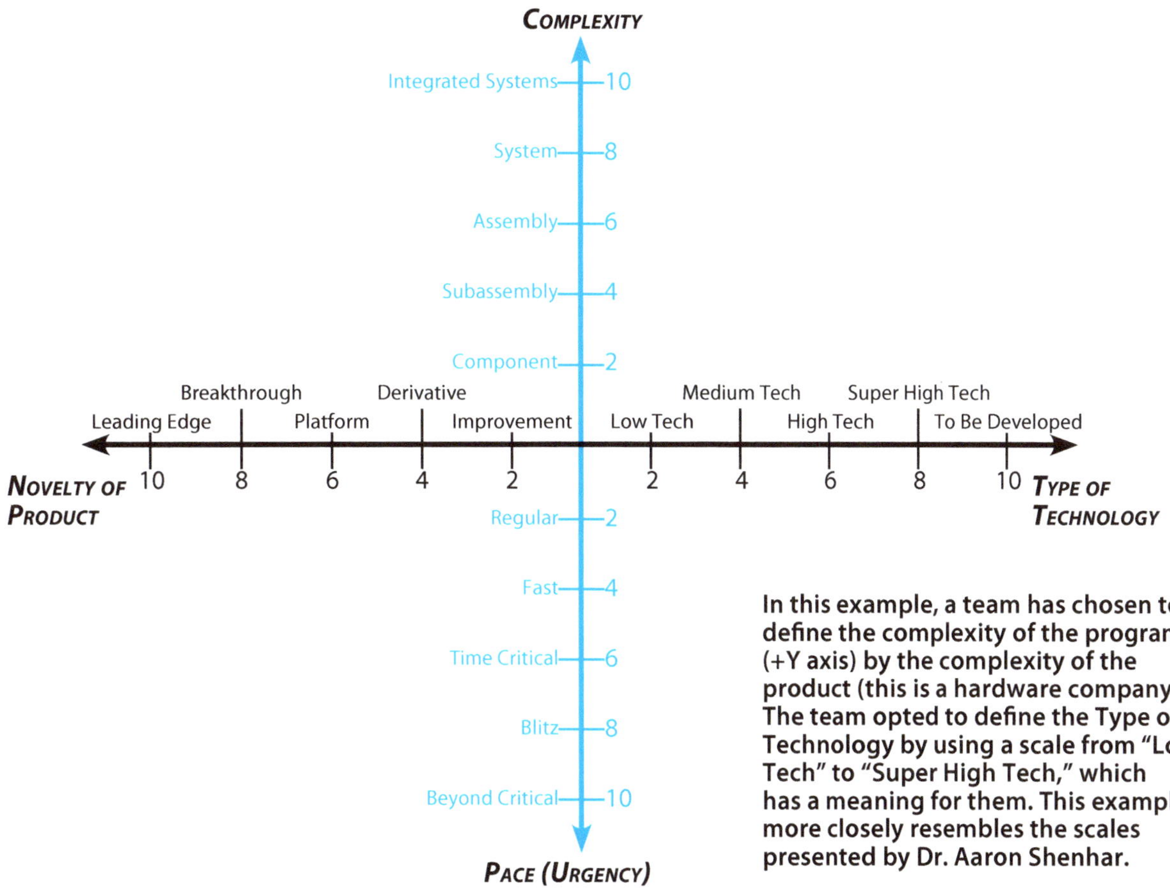

In this example, a team has chosen to define the complexity of the program (+Y axis) by the complexity of the product (this is a hardware company). The team opted to define the Type of Technology by using a scale from "Low Tech" to "Super High Tech," which has a meaning for them. This example more closely resembles the scales presented by Dr. Aaron Shenhar.

Critical Definitions and Concepts

Example 1

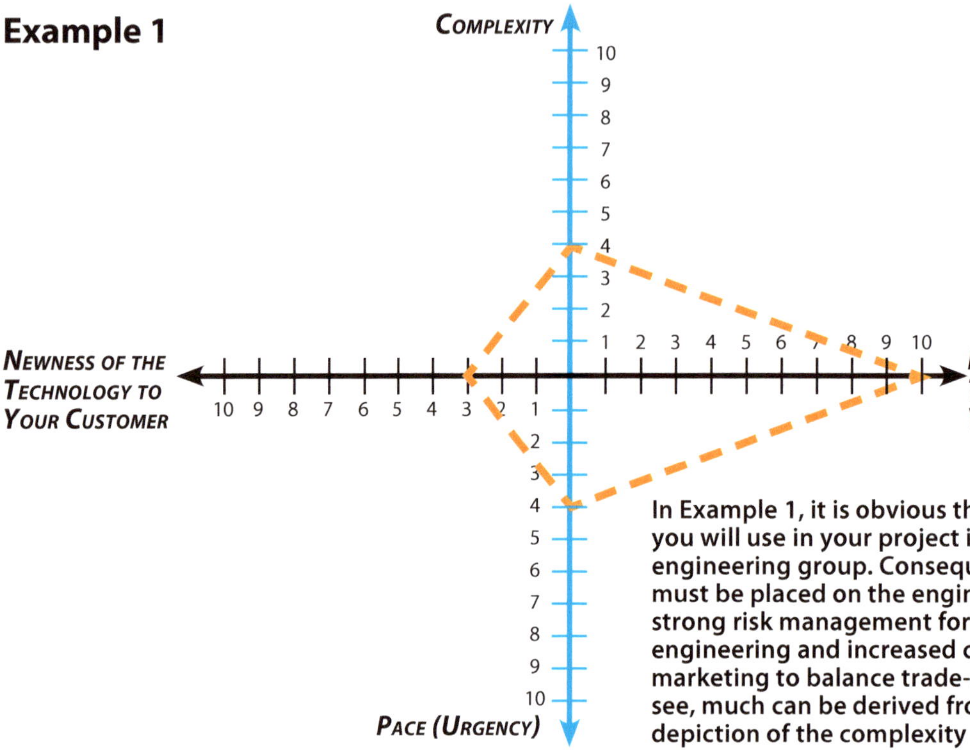

In Example 1, it is obvious that the technology you will use in your project is new to your engineering group. Consequently, special care must be placed on the engineering schedule, strong risk management for technology and engineering and increased communication with marketing to balance trade-offs. As you can see, much can be derived from having a visual depiction of the complexity of your project.

Example 2

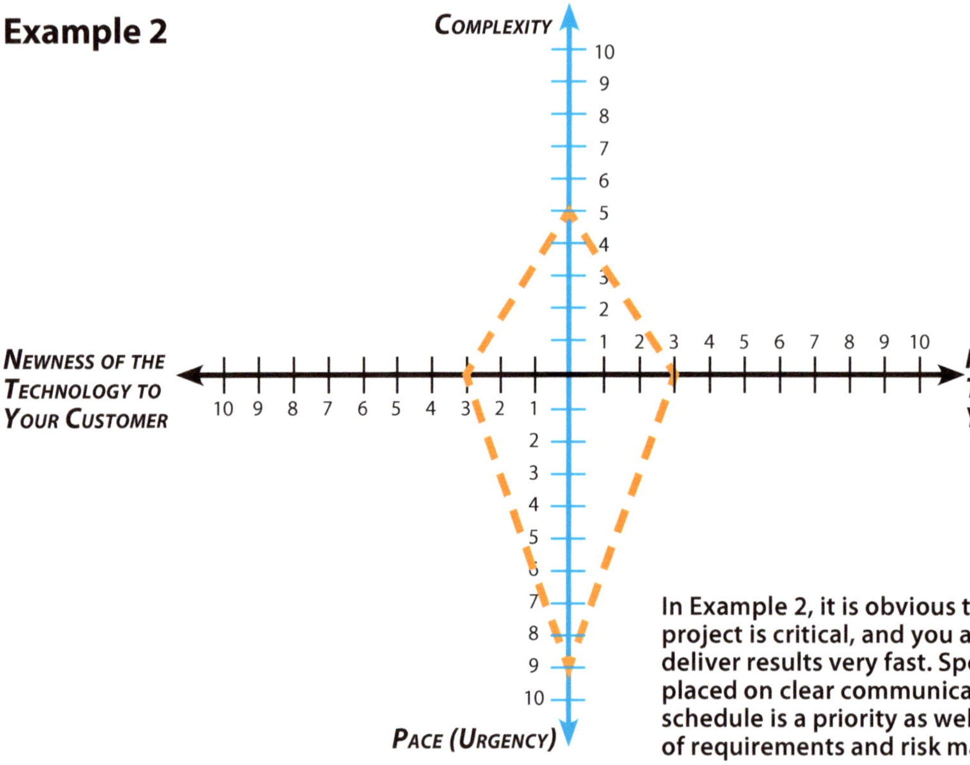

In Example 2, it is obvious that the pace or your project is critical, and you are being asked to deliver results very fast. Special care should be placed on clear communication. Tracking the schedule is a priority as well as focus on clarity of requirements and risk management..

For comments, questions or additional copies please e-mail the author: flexpm@rapidinnovation.com.

Critical Definitions and Concepts

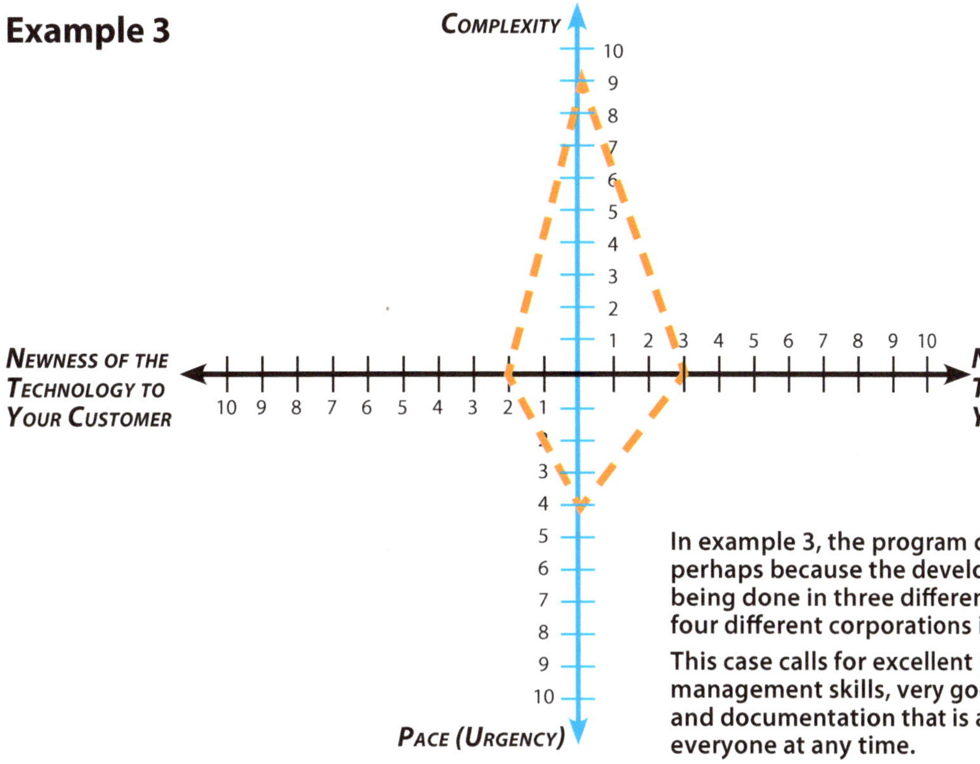

In example 3, the program complexity is critical, perhaps because the development program is being done in three different countries, with four different corporations involved.

This case calls for excellent project management skills, very good communications and documentation that is available to everyone at any time.

Developing the Scales

You can select the scales to depict the complexity of your project. Shenhar uses a 1 to 5 scale, using nouns for each of the increments between one to five. A scale of 1 to 10 is also acceptable. But remember that these are **QUALITATIVE** scales; consequently accuracy is limited, and so is resolution. Specifically, adding more resolution (granularity) does not improve the accuracy.

Generally these complexity diamonds are qualitative, not quantitative, as they are primarily a tool for communication; that is, to explain the complexity of your project to your team, your managers and other stakeholders. So be careful with too much resolution — for example, a 1 to 100 or even a 1 to 1000 scale adds little accuracy and may tend to confuse people.

You should also attach a description to each level of the scale you develop — for example, a 1 may mean "very slow and comfortable pace." If you use a 1 to 10 scale, you may want to define every odd number. With a description for each of the levels of your scale every one can understand the significance of a number.

These scales should be developed in full collaboration with your team, which will produce better scales and create support within your team.

Document your scales for future reference, and also to improve over time. Once you start using your scale you may discover new ways to improve its clarity, remain flexible.

Critical Definitions and Concepts

Detailed Process: Develop a Complexity Diamond

1

Communicate the Value of the Complexity Diamond

Before you implement make certain that all members of your team understand the Complexity Diamond and the benefits of using it to evaluate your projects.

You can use the materials in this section to communicate and persuade your team members, managers and executives.

This dialog is important to create alignment and support.

2

Obtain Clear Approval to Implement

Prior to implementation make certain that your managers and executives agree to support your efforts.

We suggest that you propose to try the Complexity Diamond on one project; that is, a pilot. This will reduce the risks and reassure your managers that it will be a gradual deployment.

3

Define Variables and Scales

Work with your team to define the four variables you will use to characterize your diamond. Ensure that the variables are fully documented for future reference.

Develop the scales by defining the meaning for each of the levels — we suggest a 1 to 10 range. Ensure that the scales and all the levels are documented.

Once you do your first project, you may need to revise the scales.

4

Identify the Pilot Project

Select a project that has not started.

Develop the Complexity Diamond in full collaboration with your team, remember that it will not be perfect the first time — but your objective is to gain experience with the tool.

Document the Complexity Diamond. We suggest you use flip chart paper or other large format such that every one can view it.

Where You're At

At this point, you have developed a Complexity Diamond for the pilot project you selected. Your team actively participated in the development process.

Continued

For comments, questions or additional copies please e-mail the author: flexpm@rapidinnovation.com.

Critical Definitions and Concepts

CONT'D

5

EVALUATE THE PROJECT

Study the values of the diamond with your team.

Reach consensus on the meaning and the implications. **THIS IS THE MOST IMPORTANT STEP IN THE PROCESS.**

Compare the evaluation with the way the project is currently organized:

- Is the competency of your team adequate to address the shape of the Complexity Diamond?
- Is the project more complex overall than you anticipated?
- Do you need to revise the schedule, budget or objectives?

6

SHARE THE RESULTS WITH MANAGERS AND EXECUTIVES

This is when the Complexity Diamond starts to produce results. You now have an objective tool that will help you communicate complexity.

Present the Complexity Diamond to your managers and executives to communicate its complexity and to share the conclusions that you and your team reached in the previous step.

This is a good opportunity to solicit resources, request changes and ask for support from the mangers.

7

CONTINUE TO EVALUATE YOUR PROJECT

After the initial characterization of your project, continue to update it as new information becomes available.

For each revision draw a new Complexity Diamond. This will help you notice the changes from your initial assumptions.

Continue to manage your project using the implications and conclusions from your Complexity Diamond.

Continue to involve your team and managers.

LEARN AND IMPROVE (KAIZEN)

At the conclusion of your project conduct a debrief (postmortem) on the Complexity Diamond.

Make changes and improvements based on the input from your team and managers.

Document the process you used so others in your organization can benefit.

Archive material for future reference, to reuse and to train others in your organization.

WHERE YOU'RE AT

The pilot is done. At this point you have used the Complexity Diamond with one project. You have also evaluated the benefits and made appropriate changes.

Now, start using it with other new projects. Remember, not all new projects require a Complexity Diamond; rather, use it selectively on projects that in your judgment may prove complex for your organization

Establish an archive so that you can re-use the definition of the scales and evaluate new projects by referring to past Complexity Diamonds.

Critical Definitions and Concepts

Product Development Frameworks

Projects don't exist in a vacuum — they interact with everything else in the organization. Also, the parameters for each project are unique. For example, the resources in one project cannot be used in another project at the same time. The same can be said about the budgets, priority and strategic importance: every development project is different.

The End-To-End Produce Life Cycle

In this series of guidebooks, we use a Whole System Map (see diagram on the next page) to explain the role of product development as part of a for-profit organization.

The system starts with the development of the corporate strategies (**T_{-4}**). This is when your executives decide the future course of the company. It's when key decisions are made that will impact the entire company for years to come. The decisions made at this point (**T_{-4}**) drive the next element in the system: the Product Roadmap (**T_{-3}**, to **T_{-1}**).

The Product Roadmap is how corporate strategies are translated into product plans. The process starts when all ideas for new products are considered (**T_{-3}**), through a sequence of decision gates, and these ideas are evaluated and prioritized. This prioritization is strategic first (**T_{-3}** to **T_{-2}**), based on competitiveness, business objectives, growth strategies, technology availability and other key variables. The tactical evaluation (**T_{-2}** to **T_{-1}**) is based on resource availability, competitiveness, market demands, etc.

Generally, once the product is approved for development (**T_{-1}**) there is a period for staging, the front-end organization of the project. This includes the fuzzy front end — that time in the project where there is confusion and more questions than answers. This phase is normal, which we believe every development project has. This is the time to obtain the answers to all the questions, more appropriately this is the project definition phase.

The official start of the project (**T_1**) is generally characterized by the Kick-Off Meeting. Not only does a formal launch communicate to the entire organization that the project is official and that is has started, but also it is a critical point to help exit the fuzzy front end. Incidently, after formal development begins, comes New Product Development Process (NPD), which we show later in this section.

The day where the product is available for customers to purchase is known as the Public Announcement date (**T_2**). Please note that many companies may define the PA date differently; for example, when the product is ready to ship to customers, or when the product appears in the trade press.

During the product development process (**T_1** to **T_2**), your company is spending resources, in many cases in the millions of dollars. This is the investment your company makes in your new product and with the expectation of a return on that investment (ROI). After Public Announcement (**T_2**), your company starts selling the product and recovering the investment until it matches the revenue from sales (**T_3**). After that point your company finally starts making a profit.

Time-to-profit is the point that your company has set to measure the success of your new product. It is a point set prior to the start of the project (**T_{-1}**). This is a critical metric as it defines the business success of the new product.

Critical Definitions and Concepts

Process Map: Project Management and the Product Life Cycle

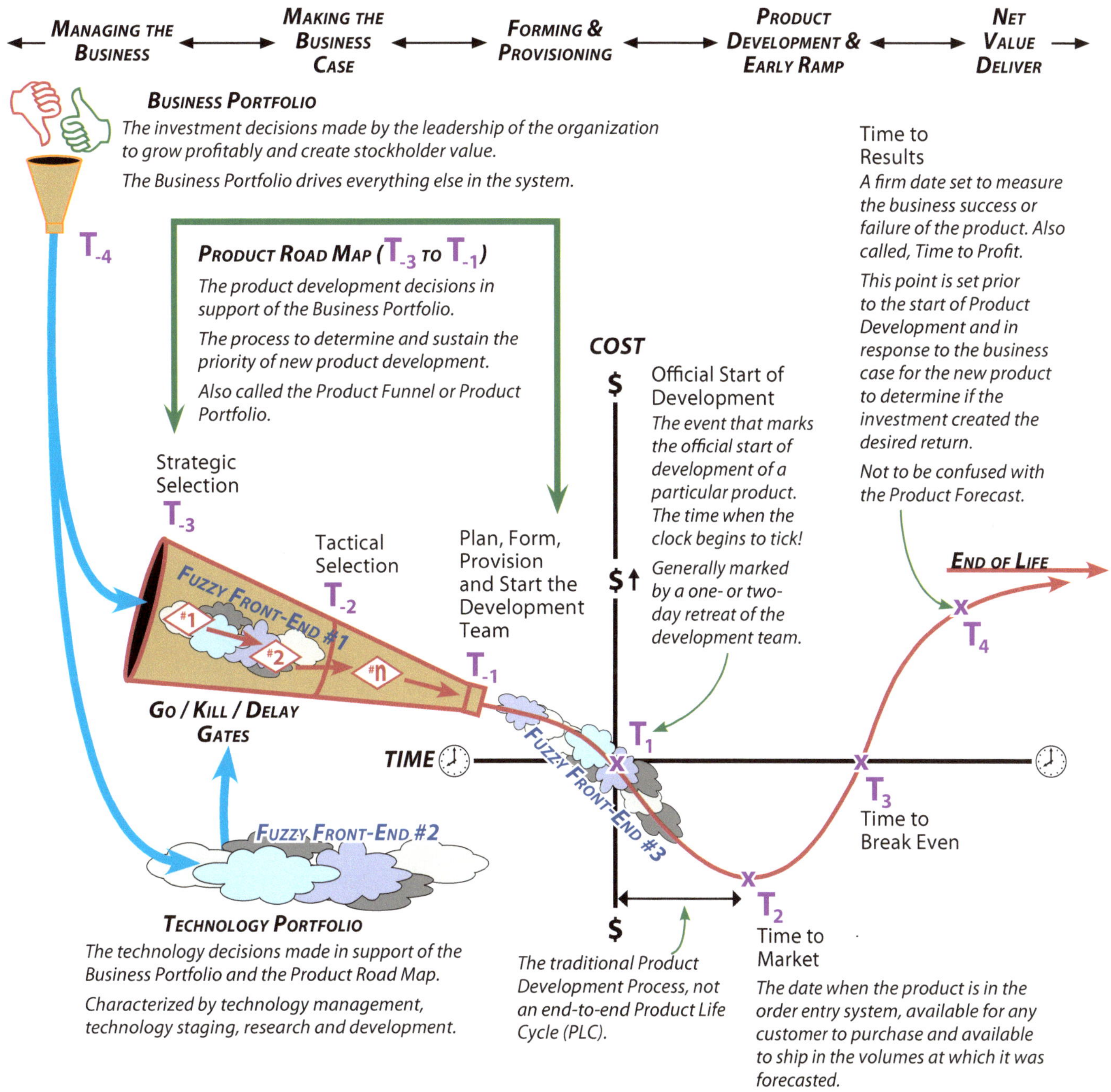

Important

1. This is a conceptual model, not drawn to scale.
2. Other elements that are very important but not detailed in this diagram include the manufacturing process, end-of-life (also called phase out or cash out), a system of metrics and a feedback loop and the role of the customer (which is so critical to the above diagram, it deserves its own model).
3. The system is adaptive and flexes based on competitiveness and other business variables.
4. Only one product development curve is depicted. Concurrency is implied; that is when several products are developed at the same time.

Critical Definitions and Concepts

Life Cycle Models

New Product Development Processes (NPDs)

There are many processes that have been created to develop new products, all aimed at achieving three goals:

1. Reduce time to market.
2. Meet customer requirements.
3. Achieve profitability goals.

The journey to improve product development continues as the world changes. Your company probably already has some sort of product development process.

Major Categories of Product Development Processes

Method	Description
1. Ad Hoc	No process or framework; the product development process is created as the development progresses. Not desirable as it does not provide the opportunity for continuous improvement. It is the most unproductive of all approaches.
2. Waterfall Model	A software development model first proposed by W. W. Royce in 1970. Royce presented it as a flawed, non-working model. It's a sequential development model in which each phase is completed before the next one starts; thus, like a series of waterfalls, from requirements, to design to implementation, to testing and integration. This model is similar to the Phase Gate model for hardware development. Modern software development is much more dynamic than the Waterfall model allows; consequently it has fallen into disfavor.
3. Spiral Model	Developed by Barry Boehm to address the weaknesses of the Waterfall model. It focuses on addressing risks incrementally by repeating the Waterfall model in a series of cycles or rounds. It provides for multiple builds and several opportunities to meet customer requirements.
4. Iterative	Very similar to the Spiral model, and the start of the Agile approach for software development. Extreme Programming is another member of this category. Iterative development is a rework scheduling strategy in which time is set aside to revise and improve parts of the system.
5. Incremental	This model involves developing various part of the product at different rates and then integrating them as they are completed. Iterative and Incremental were merged in the 1990s through the Rational Unified Process (RUP).
6. Agile	The Agile approach promotes the development interactions through the life-cycle of the development project. Characterized by developing software in short amounts of time from one to four weeks, where each interaction is defined as a software project. This model requires or assumes constant access to customers and is best suited for IT environments.

Critical Definitions and Concepts

Method	Description
7. Concurrent Engineering	A methodology that recognizes that in product development there are opportunities to develop various parts at the same time — i.e., no need to wait for one part of the product to be defined before staring another one. Well suited for rapid product development, it allows a Core Team to serve as the steering mechanism for development. The Core Team is made up of representatives of the major functions involved in the development such as engineering, marketing, manufacturing, service, finance, etc.
8. Phase Gate	Mostly used in hardware development, it involves a series of review gates along the product development cycle. These gates are generally associated with the major milestones of the project. A very common process in the hardware world, but not sufficient for fast-changing markets where speed is essential.

More on the Product Development Process

All product development processes have one objective in common: Deliver a profitable product to the market from idea to saleability. It is an absolute that all new products start with an idea — perhaps not even noticed, but an idea nonetheless. From this point, there are may approaches and methods to reach completion (as you can see in the next table). Another absolute is the need to create value and deliver it so that your customers will pay and your company will make a profit. As a project manager you must always have these absolutes in mind; they will help you lead your team to success. Whatever methodology you select, from Agile to Waterfall, the need for dealing with it as a project does not change.

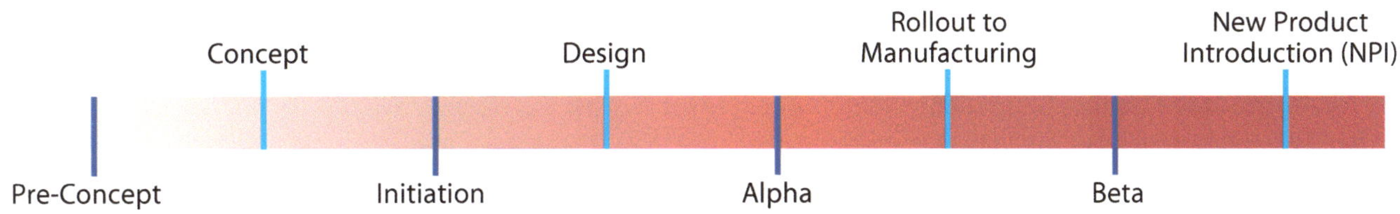

Software Development

Software development differs from hardware in several ways, but the need to deliver value to the customer does not change. The authors recognize that the processes do have differences, but at the core they are very similar. Their implementation may differ, but at the strategic level they are very similar.

If anything, newer software development practices require focus on project management. In the end, the need to clearly understand customer requirements beyond technical needs and translating these needs into new products is as important for software as it is for hardware. Both provide a solution to the customer's problem, and usually the solution is not just technical, regardless of software and hardware.

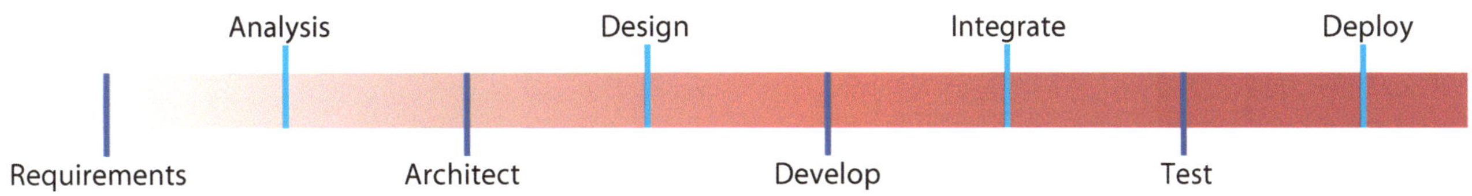

Critical Definitions and Concepts

The Bottom Line

There are yet more product development models than those we've highlighted so far, and new ones are emerging as well. The key message is that you create or adapt a model to suit your business objectives and the culture of your organization.

Remember that every product development process has three objectives: reduce time to market, meet customer requirements and achieve your profitability goals. It's OK to experiment, but you and your team must clearly understand that it is an experiment and you must have a clear reason for the experiment. Don't be fooled by new fads and models which promise much and deliver little.

The best way to improve your product development process is to have formal debriefs (postmortems) and a robust process to deploy the improvements. A steady flow of debriefs will yield considerable improvements and it has shown to be the best way to improve the productivity of the your development team.

Flexible Project Management and NPDs

Project Management practices and methods apply regardless of the NPD process selected. Remember that regardless of the NPD process, all product development endeavors are projects, by any standard. Consequently you can feel comfortable with applying project management methods to any product development.

The Bounding Box

An Effective Tool for Accelerating Your Development Project

Complexity in product development can be reduced through the use of the Bounding Box , also called the Strike Zone. It improves communication and, particularly, decision making. As the name implies, it is a set of boundaries given to a development team. As long as the development program stays within the established boundaries, the team is free to make all decisions.

The boundary conditions are negotiated between management and the development team, specifically the Core Team. These boundary conditions can include budget, timeline, profitability in the form of margin or manufacturing cost and other key elements. Naturally the list of boundaries must be finite and limited to critical performance factors. The concept is not dissimilar to MBOs (Management by Objectives) and KPIs (Key Performance Indicators) except that it is project specific and it involves the development team.

In addition to agreeing on the key parameters for the development project, a communication and escalation protocol must be developed. The development team and management must agree on a communication protocol that allows all parties to feel informed but with minimal micro-management. The escalation protocol declares when the team must consult with management; for example, if one of the boundary conditions will be exceeded.

The Bounding Box is challenging for organizations where there is no cultural history of empowerment, as management is releasing some control. But this act of empowerment is a worthwhile investment because it accelerates product development.

Benefits of the Bounding Box

One advantage of the escalation method is that as long as the project in on track, the team is free to manage itself with reduced top-down micro-management. This eliminates needless meetings and provides the development team the flexibility to act efficiently. Another advantage is that decisions are made at the closest possible level to where the work is being done.

Essential

"The Bounding Box has been shown to reduce time-to-market and reduce complexity."

Critical Definitions and Concepts

When To Use a Bounding Box

Generally, here are the reasons for introducing the bounding box tool:

- Program complexity
  - Many internal and external factors are critical to success.
  - Need better tools for simplifying, communicating and focusing on the critical success factors.
- Dynamic and unpredictable environments.
  - Using the best practices and lessons learned from past projects is important but not sufficient.
  - Need better tools for anticipating change and quickly adapting to it.
- Time-to-market — especially in the early phases.
  - Current management tools work best at controlling after objectives are set and plans are in place.
  - Need better tools for managing before teams and management have all the information.

Note: Not all projects must have a bounding box; It is better to save the model for projects that are critical to the success of the organization.

What to Include in Your Bounding Box

It depends; this is work that must be done in close collaboration with the executives of your organization and the Core Team.

Remember, the value of your bounding box is in its simplicity. Don't go overboard with the components. Use the fewest items that will enable the team to move quickly while also keeping the executives informed.

In General

Generally, the boundaries or bounds are defined based on the following parameters:

- Strategic importance to the organization — i.e., how important is the project to the future of the organization?
- Program complexity — i.e., small, medium and large.
- Critical success factors — i.e., time-to-market, hardware and/or software product costs, overall development cost, market size, etc.
- Uncertainty of the environment — i.e., technical, market, political, etc.
- Need for clear communication between the team and the executives of the organization.

Definition

CORE TEAM

Small group chartered and empowered to guide the implementation and completion of an initiative.

Very useful in fast-changing environments where distractions abound. A small group can focus attention and energy to ensure clean execution.

The roots of Core Teams can be traced to methods of Simultaneous Engineering and Concurrent Engineering.

More Specifically

Specific items that might be important to include in your bounding box

- Achievement of the value proposition — for example, the critical new features that will ensure success in the marketplace.
- Definition of the target market — for example, the specific segment that your new product will serve.
- Technology strategy.
- Technology milestones.
- The very few critical risk factors.

Critical Definitions and Concepts

- Financial limits — for example, the amount of investment in the new product, target selling prices of the new product, sales in Year One and profits in Year One.
- Critical requirements for execution — for example, tools, training, staffing and skills.
- Global considerations.
- Go to market criteria — that is, the short list of business and technical requirements without which you will not roll out the product to the marketplace.

The Criteria for Your Bounding Box

- The bounds reflect the top few ingredients for success of your program — i.e., quality over quantity.
- The bounds are objective and measurable.
- The escalation process is fully documented and agreed to by the executives and the team. It must include the bounds, roles and responsibilities, frequency of reporting and criteria for escalation. Pay special attention to clarifying the trigger points; that is, events that will trigger an escalation.
- The bounding box is created at the start of your program, at the time you launch the development team.
- The bounds are specific to your program. They are *not* generic and *not* one-size-fits-all.
- The definitions of "in-bounds" and "out-of-bounds" must be clear, measurable and fully agreed to.

Empower Development Teams

- Within these boundaries, the team and management agree that the program is on-track.
- Outside the boundaries, the team escalates issues or decisions to management.
- As long as the program is in-bounds and within the zone, the team makes the day-to-day decisions and adjustments.

Critical Definitions and Concepts

Example: The Bounding Box as a Proposal

Start

Alamo Proposal #1

FINANCIAL PERFORMANCE

Calculate following metrics at least every other AP:

- Max. Total NRE Cash Flow: $1.2M
- Min Total Net Sales: $35M
- Min Gross Margin at Product Ship Release: 70%
- Max. Capital Cost: $150K

	MCOS GOAL	MCOS ACTUAL (11%)
v(15)	$15,300	
v(13)	$12,750	
x6	$4,200	
x7	$13,800	

CUSTOMER DRIVEN MILESTONES AND SCHEDULE

- Public Announcement: .... TBD by 08-31-01
- Availability for BB Product: .. TBD by 08-31-01

PRODUCT FEATURES AND FUNCTIONALITY

- 1300nm and 1550nm Laser versions
- External Tunable laser input
- 2 Receive Data Electrical Outputs

INTERNAL MILESTONES AND SCHEDULE

- Not to Exceed Product Ship Release: .... March 2005

Nine Months Later

Alamo Proposal #2

FINANCIAL PERFORMANCE

Calculate following metrics at least every other AP:

- Max. Total NRE Cash Flow: $1.6M
- Min Total Net Sales: $20M
- Min. Gross Margin at Product Ship Release: 70%
- Max. Capital Cost: $150K
- Min. Profitability Index (with Sunk Cost) ... 2.7

	MCOS GOAL	MCOS ACTUAL (11%)
v(15)	$15,850	$15,787
v(13)	$13,300	$10,146
x6	$5,500	$5,342
x7	$14,300	$13,685

CUSTOMER DRIVEN MILESTONES AND SCHEDULE

- Public Announce: .. 30 days before June 2002
- Availability for BB Product: .... June 2002

PRODUCT FEATURES AND FUNCTIONALITY

- 1310nm and 1550nm Laser versions
- External Tunable laser input
- 2 Receive Data Electrical Outputs

INTERNAL MILESTONES AND SCHEDULE

- Not to Exceed Product Ship Release: .... June 2005

This example is from an actual project, with a few changes to protect confidential information.

Notice that the parameters are limited to a few critical elements of the development program; thus, keeping the model simple to manage. A green dot indicates that the project is within bounds, everything is normal. The first proposal (left) shows the starting projections.

The next proposal (right) shows the project nine months after initiation. Red text indicates what's been altered. A red dot indicates a change since the last project review, resulting in the escalation protocol being put into motion; for example, the team comes before management to discuss alternatives and find the proper corrective action.

Critical Definitions and Concepts

Example: The Bounding Box as a Visual

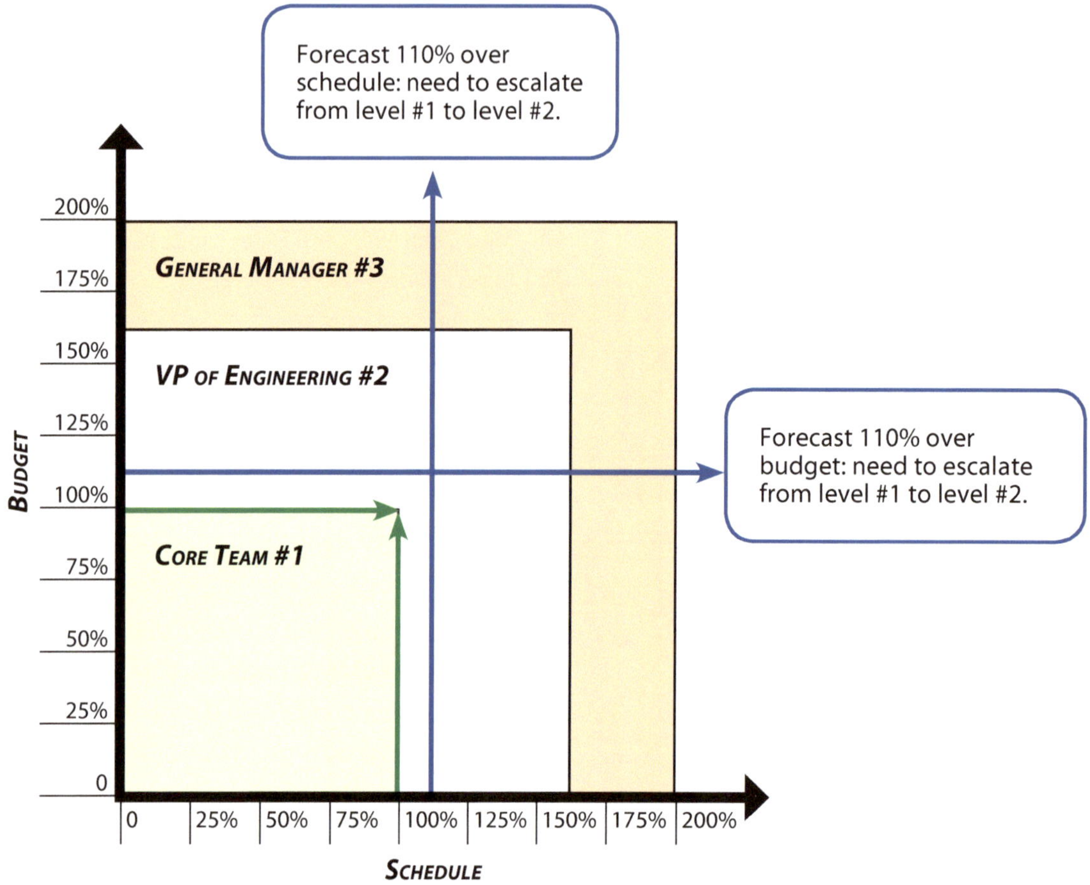

This visual example of the Bounding Box can be used to communicate with the entire team, managers and executives of the organization. Plotting the escalation thresholds is also a good way to empower the team to make its own decisions.

Notice the vertical and horizontal blue lines that track the budget and the schedule. As long as the two lines are within the inner square (#1) the team is free to make decisions. If either line falls outside the inner green square, then the issue must be taken to the next appropriate level of management (#2) or (#3). These are not dependent variables, but simply relational positions; that is, one line is not a function of the other. Also note that the tracking must be done regularly, and you need to monitor the position of the two lines.

In the above example, we see that the project is 110% over budget (blue horizontal ines); consequently, it must be taken to the next level of escalation, in this case the VP of Engineering. Notice that the schedule line is still within the Core Team (green area). So no action is needed.

NEW TOPIC

Leadership

Due to the complexity of product development, effective, flexible leadership is critical. We strongly suggest that you develop your leadership capabilities by reading books on the topic, researching it online and attending seminars and workshops.

Leadership is becoming yet more critical due to the advent of multi-site product development, where a mix of cultures is present, not to mention languages, ethics, social structures, etc. We also suggest that you learn about leading multi-ethnic teams. There are several books on the topic, and new ones are always being introduced. Search the Web for information about multi-ethnic, virtual teams, and multi-site product development.

In its simplest definition a leader is one who has followers. A more complex definition might be your ability to enable your team to succeed through your actions, language, vision and convictions. It is your ability to guide your team from the start to the successful delivery of a profitable new product.

Comparing the Ineffective Leader to the Effective Leader

The Ineffective Traditional Leader	The Effective Flexible Leader
1. Keeps a strict routine.	Gets rid of useless habits and hates bureaucracy.
2. Maintains.	Improves and develops.
3. Does it by the book.	Open to ideas from anybody.
4. Focus on Structure.	Focus on people and teams.
5. Commands and controls.	Inspires.
6. Short term view.	Long term view.
7. Imitates.	Innovates.
8. Fixes blame.	Learns from mistakes.
9. Inspires fear.	Empowers and encourages.
10. Plans and organizes.	Communicates, consults, coaches and mentors.
11. Provides all the answers.	Asks the right questions.
12. Builds dependence.	Creates new leaders.

In stable environments the attributes of a traditional manager may be sufficient to lead. But in product development, where change is constant and many variables impact your development project, the need to be an effective, flexible leader is critical.

Evaluate your style of leading using the table above. Are you an effective leader? Are you a flexible leader? What do you need to change to become a better leader?

Advanced Communication

The Power of Effective Communication in Product Development.

Important: The traditional status report is woefully insufficient to ensure communication.

The more chaotic your environment, the more you must rely on clear communication because the traditional project management tools lose their effectiveness. People are constantly bombarded with a myriad of messages, all competing for attention, and yours is but one of many. Communication skills are critical to your ability to lead a development project in a turbulent environment. We encourage you to constantly work to improve your ability to communicate.

Develop a plan to improve your ability to communicate and set about implementing it. This guidebook is not sufficient to meet the level of communication you need to effectively lead a development project. However, it does provide several tools and methods that can improve your ability to communicate. We strongly suggest that you seek additional training in the form of reading, seminars, formal education and other means to continually hone your skills.

Traditional communication is not sufficient in a chaotic environment. Usually, project managers communicate through status reports, which are written documents with information about the status of the project. Caution! If you rely solely on a status report, you are falling short of communicating the progress of your development project. Assume that the status report is only 10 percent of the communication you need to deliver. Furthermore, sending an email with the status of your project does not constitute good communication. Besides, people are flooded with too many emails as it is.

Others rely on project review meetings, where the project is reviewed by the project manager and key stakeholders are invited. These traditional methods do not provide the coverage nor the engagement to ensure clear communication. People are inundated with information and do not have the time to clearly understand the messages. It's as if you are shouting to them across a chaotic, noisy room.

Smart project managers in product development also add a second layer of informal communication such as conversations and informal meetings. In addition, the use of web pages and other electronic communication help. But even that is not enough in chaotic environments.

One-on-one meetings with key personnel are essential, as are bulletin boards in highly trafficked areas and war-rooms — areas that are constantly updated to keep every one informed.

Attributes of Effective Communication

Clarity is essential. The information you send must not be misinterpreted. Write down the points you want to communicate. And then review what you've written! Check for spelling and grammar as well. Effective communication must be to the point, in the vernacular of the reader and convenient.

Stay on topic. Know what you are trying to say. The old journalism formula is a wise way to communicate: **WHO, WHAT, WHEN, WHERE AND WHY**. An action item must clearly explain what needs to be done, by whom and by when; otherwise, you don't have an action item.

Your audience will not likely see the world the same way you do. Consider that each person has a different function in the project, let alone perspective and set of priorities. Write your information in their language, not yours, being sensitive to jargon, acronyms and idioms. Be cognizant of how formal or casual you want your communication to be. Pay special attention when communicating with those who speak a different language than you or were raised in a different culture.

Make it convenient for your reader to access information. For example, while email may be great for you, it may not be the best way to inform others. Remember, convenience is only so when the receiver says it is.

Managers and executives are always interested in any changes since your previous communication, any new obstacles or roadblocks, what will happen next in your project and any requests for their support regarding your product.

Multi-Ethnic Teams and Communication

Product development is now global. The improvements on telecommunications, air transportation and technology in general have enabled companies to vastly expand the approach to product development. It is not rare to have an engineering team made up of people from three and even four different countries. This is a significant business opportunities for all, but at the same time creates new challenges when trying to communicate effectively.

It has always been true that employees who live farthest from the head office lack the most information about projects. Imagine a development team made up of engineers in Shanghai, China, Kiev, The Ukraine, and San Jose California. There are at least three different languages involved, not to mention different ethnic cultures, and probably different corporate cultures — all very serious challenges to clear communications.

Recommendations

- We first recommend that you recognize there are obstacles to good communication in multi-ethnic, multi-site development teams. Alert, you can avoid such obstacles and also search for better ways to communicate.
- We also suggest that you develop a formal communication plan, where you carefully evaluate the special needs of the team and take steps to overcome barriers. Focus not only on technological needs, but also on cultural, language, time and place needs. For example, English may be a second language to most of your team; consequently, you need to take into consideration this barrier when writing your MRD (Marketing Requirements Document) — while the language may seem very clear to you, it may be a daunting challenge to others.
- We strongly recommend that you travel to the countries where the rest of your team is located. It may be costly, but the consequences of not traveling are even more expensive and show up in delays, defects, and poor implementations.
- Your team members from other countries should attend your Kick-Off Meeting. This will accelerate your development project, strengthen your team, and significantly improve overall communication. Do not let the expense interfere with your need to do the right thing.

Important

Your team members from other countries should attend your Kick-Off Meeting — the authors believe that this is non-negotiable.

Important Note About Technology

Technology has indeed enabled the growth of multi-site product development. The Web, video-conferences, peer-to-peer video, email and many other tools have helped the transfer of information. Yet, these same technologies can give you a false sense of communication. For example, just because you conduct video-conferences does not mean that you are communicating effectively? There are obstacles to communication such as culture, language, distance and even individual styles. We encourage you to constantly improve your ability to communicate effectively — don't rest because you have a particular technology.

The Value of a Communication Plan

Clear communication is essential to successful project management! Yet it is one of the more neglected duties in project management. Lack of communication leads to waste, friction and much aggravation. It the responsibility of every project manager to keep everyone informed, from the project mentor to the project team to the project customers and beyond.

Clear communication is not a casual affair! We live in an hyper-communicating world; thus, the challenge for project managers is to break through all the noise and deliver the right message to the right person in the right form and at the right time. Furthermore, we tend to underestimate the need for

Essential

"The challenge for project managers is to break through all the noise and deliver the right message to the right person in the right form and at the right time."

Advanced Communication

communication, perhaps due to our familiarity with the project, which makes us assume that everyone else knows about it.

We also tend to underestimate the needs of team members who reside in other locations. But that distance aggravates the lack of communication. You add to this different languages and cultures, and the stage is set for delays, errors and waste due to poor communication.

Questions to Ask When Developing a Communication Plan

- Whose support and cooperation do we need?
- What organizational conflicts exist between stakeholders? And what do we need to do to anticipate trouble?
- What lines of communication exist among interfaces?
- How can we improve communication across functional boundaries?
- Who are the key decision-makers for this project? And what information do they need?
- Whose opposition should we expect? And what can we do to preempt any friction?
- Who are the project customers? And what information do they need?
- Who are the project stakeholders? And what information do they need?
- Who are the members of the project team? And what information do they need?
- Who are the contractors, consultants and other outside resources? And what information do they need?
- Are there any members who are not collocated? What are their informational needs?
- Have we addressed the communication needs of everyone in our responsibility matrix?
- How do we take full advantage of project meetings, status reports and other meetings to improve communication?

Key Elements of your Communication Plan

- What needs to be communicated? (What type of information do they need?)
- Who needs to receive what information?
- When do they need the information?
- Who in our team will be responsible to ensure that we deliver clear information?
- What format will be used for the various types of information? Are these formats convenient to the addressee?
- What information must be delivered in person?
- How often will each constituency need to be informed?
- How will we assess the effectiveness of our communication?

Examples: Communication Matrix

The 2x2 Matrix is one way to prioritize your communication efforts. Use the 2x2 Matrix tool to map various constituencies pertaining to your project by assigning them to a specific quadrant. You can map individuals, groups, department or specific functions to categories such as influence on the project, interest in the project or understanding of the project. The process will reveal to you the amount, frequency and energy-level of the communication you will need to put forth to keep your constituencies informed.

Template

See Also: Critical Tools: Karnaugh Maps

Thoughtfully determine what two criteria will comprise the matrix. Take time to consider in which quadrant each constituent best fits. This is not a casual activity, as it will define your communication strategy. Be sure to map with enough specificity to allow you to direct communication effectively.

Example 1: Abstract

In this example, the upper right quadrant is the most important. It's where the people with the most interest and the most power to influence are located.

Where would you place your team members?

They have high interest and high power to influence because they will do all the work. Consequently, we would map them in the upper right quadrant.

Vertical axis: POWER TO INFLUENCE YOUR PROJECT (0–10); horizontal axis: LEVEL OF INTEREST IN YOUR PROJECT (0–10)

HIGH INFLUENCE LOW INTEREST What to do: Understand their needs and keep them satisfied.	**HIGH INFLUENCE HIGH INTEREST** What to do: Keep these key players delighted.
What to do: Minimal effort; monitor occasionally. **LOW INFLUENCE LOW INTEREST**	What to do: Keep informed, identify any concerns. **LOW INFLUENCE HIGH INTEREST**

Example 2: Concrete

This hypothetical example shows the communication strategy for a development program. You will need to keep the list available and review it frequently in order to keep the right people informed with the right information.

Make your communication convenient to the individual. Ask each person what is his or her preferred way to communicate. Email is less effective because people receive too much of it. Face-to-face is still the ideal approach. You can also have project review and other meetings to ensure clarity of communication

Vertical axis: POWER TO INFLUENCE YOUR PROJECT (0–10); horizontal axis: LEVEL OF INTEREST IN YOUR PROJECT (0–10)

HIGH INFLUENCE LOW INTEREST Ming Cho: Quality and Reliability Sophie Lui: Service and Repair	**HIGH INFLUENCE HIGH INTEREST** Fred Smith: General Manager Mary Moore: VP of Engineering Lucy Rogers: Product Manager Hung Far: Manufacturing Manager
Central Engineering Organization Operations Manager Logistics Manager **LOW INFLUENCE LOW INTEREST**	Development Team Members Jose Mendes: Chief Technology Officer Jean-Claude Balland: Sales Manager **LOW INFLUENCE HIGH INTEREST**

Advanced Communication

Example: Communication Decision-Making Tree

Project Name: "LARGO" Date: May 2009

Project Leader: Gary Smith, Senior Design Engineer

Project Sponsor or Mentor

Fred Tang
VP of Marketing

Key Decision Maker

(A) Doreen Woo
Division General Manager

Other Key Stakeholders

(R) Charles Mendoza
Engineering Manager

(R) Sarah Gomez
Marketing Manager

Project Team Members

Name: Charlie Roof
Title/Function: Mechanical Eng.
Contact Info: Charlie@xyz.com

Name: Mary Marino
Title/Function: Sr. Software Eng.
Contact Info: Mary@xyz.com

Name: Tami Carloni
Title/Function: Reliability Eng.
Contact Info: Tami@xyz.com

Name: Jon Marshall
Title/Function: Software Eng.
Contact Info: Jon@xyz.com

Name: ____
Title/Function: ____
Contact Info: ____

Name: ____
Title/Function: ____
Contact Info: ____

Name: ____
Title/Function: ____
Contact Info: ____

Name: ____
Title/Function: ____
Contact Info: ____

Name: ____
Title/Function: ____
Contact Info: ____

Name: ____
Title/Function: ____
Contact Info: ____

Project Customers

- Those who will derive value or from the project.
- Those who will purchase the end result or your project.
- Those who benefit from your project.
- Those who will use what your program delivers.

(IN) Design Engineers in small manufacturing facilities

(IN) Lead technician, manufacturing line

(IN) Assembly technicians manufacturing line

Quality Manager Manufacturing line

Code	Responsibility	Code	Responsibility	Code	Responsibility
A	Approve	C	Create	IN	Provide Input
R	Review	D	Deliver	N	Notified
		I	Inform	L	Leads a Team or Sub-Team

This template offers you another way to implement your communication plan using the Decision-Making Tree approach. It relies on the organizational structure. You enter in specific names titles and then use the code letters to define the type, frequency and extent of your communication effort.

Marketing Your Project

Marketing your project includes creating awareness and favorable opinion about your project from the various constituencies such as your mentor, key stakeholders and project customers.

This should not be confused with the negative image of marketing — i.e., misinforming and misleading. Instead, there is a need for everyone involved in your project to be informed and to form a positive image of your efforts. After all, these people can be motivated to support your project.

The Levels Of Marketing

For every project, there are various levels of awareness. There are those who know close to nothing about your project, others are well informed, while others are fully committed to providing support. And there are those who've gone beyond commitment and are already providing full support. Your challenge is to ensure that everyone is at the proper level of awareness. For example, someone may have little knowledge of your project, when in reality that person should be fully committed to helping you.

See Also
Critical Tools: Awareness Curve

For people to be able to take action to help, they must go through all the levels of awareness. First, they need to be aware of your project. Then they need to be informed. Third comes commitment — naturally, no one will commit if they are not even aware of the project. After commitment comes action, and after that, the results. Your challenge is to create a marketing plan that ensures that everyone is at the right level.

Objections

Most projects cause change of some type. At times, some will resist this change and will create obstacles that will impede progress of your project. Yet there are others who honestly have some legitimate issues with your project. The challenge is, how do you tell them apart?

Objection to change often takes the form of denial — for example, someone says, "We tried that, it didn't work," or "Our organization is different, it won't work here," or "Let's study it some more." Statements like these are known as push-back and are a possible sign of denial. Many times, people are simply looking for more information, and they manifest their interest by being negative. You need to ensure that people are receiving the right information at the right time and in the right form in order to prevent push-back from those who are merely seeking additional information.

Regrettably, some simply don't like change. They will have objections, obstacles and all types of barriers to avoid the change that your project brings. What to do? First, ensure that these persons have all the information. Then, you may have to develop an intervention — for example, recruit your human resources person to deal with the issue. In small companies, you may have to go to your manager or even the owner of the company.

Develop Your Marketing Plan

Your project communication plan and your project marketing plan are two different things; although, they do have similarities. The key difference is the intent; that is, what impression do you wish to create after the individual receives your information?

Marketing starts by identifying who you need to influence. Next, proceed to creating awareness of your project and you can achieve this through your status reports and other communication efforts. Then you need to create preference; that is, a positive image of your project (here is where your communication and marketing plans differ). Incidentally, creating a positive image doesn't mean to mislead or deceive. It simply means you should present your project based on the benefits it will create such as the achievements, outcomes and results.

Use the right tools for the marketing job. For example, you can make presentations and updates. You can put information in your Intranet and other Web-based sites. Bulletin boards, posters and signs are also part of your marketing toolbox. Use your imagination and be clever about providing the right information, at the right time.

Advanced Communication

Advanced Communication

See Also
Critical Tools: Awareness Curve

We cannot stress enough the importance of the role of clear communication in a product development environment. Traditional communication tools such as the status reports or email do not survive the noise that is ever-present in new product development. It is your responsibility as the project manager to take every measure possible to ensure clear communication, including taking personal responsibility for creating ways and means for communication well beyond the traditional set.

Please see the Critical Tools section in this book for important concepts to improve your ability to communicate and also help your team communicate more effectively.

Conflict and Conflict Management

See Also
Critical Tools: Conflict and Conflict Management

Conflict in product development is inevitable due to the passion that all participants have for the success of the development program. Also, there are always personal, departmental and other sources of friction.

You should be prepared to facilitate the resolution of conflict as the product manager as you are objectively focused on delivering measurable results. If the conflict is too complex, you may need to hire or secure outside help.

Please see the section on Critical Tools in this book for guidance to resolving conflict.

Team Building and Teamwork

Definition of Team

As the project manager, it's in your best interest to ensure that you enable the formation of a team. While most of the participants in your development program may know each other and may have even worked together before, the new development project you lead requires the creation of a sense of team. This effort should start even prior to the official start of the development program. Furthermore, we highly recommend that you hold an official kick off meeting, where your focus is to initiate the formation of the team. Use this opportunity to develop a common purpose, a common set of measurable outcomes, etc.

Teams in business environments are expected to deliver measurable results; consequently, we have selected the definition created by Katzenback and Smith as our chosen expression of the essence of a team.

> **What is a High Performance Team?**
>
> A small number of people — with complementary skills, who are committed to a common purpose, performance goals and approach for which they hold themselves mutually accountable — and are deeply committed to one another's personal growth and success
>
> —Katzenbach and Smith

The above definition is one of our favorites. It is simple, clear and applies very well to business environments where all teams are expected to deliver measurable results. The above definition was extracted from the book by Katzenbach and Smith The Wisdom of Team, which we highly recommend to further your understanding of teams in the business world.

> **What is a High Performance Team? — Remix**
>
> A small number of people — with complementary skills and who are committed to:
>
> 1. A common purpose.
> 2. The same performance goals.
> 3. A common approach.
> 4. For which they hold themselves mutually accountable.
> 5. And are deeply committed to one another's personal growth and success.
>
> —Katzenbach and Smith

We have distilled the key components from the definition of team down to five essential ingredients to make a team. We can conclude that unless you truly have the above ingredients, you do not have a team. With one exception, we suggest that the fifth criterion be reserved for high performance teams, as it is difficult to achieve. Do strive to achieving items one through four takes — it takes time and effort. It takes more than a PowerPoint presentation to make a team!

Key Ingredients for a High Performance Team

- People with complementary skills; i.e., cross-functional.
- Commitment to:
  - **COMMON PURPOSE** = Team Charter, objectives, deliverables.
  - **PERFORMANCE GOALS** = Measurable Objectives, tangible results.
  - **COMMON APPROACH** = Team Plan, Work Breakdown Structure (WBS), visual road map.
  - **MUTUAL ACCOUNTABILITY** = Team values, roles and responsibilities.
  - **ONE ANOTHER'S PERSONAL GROWTH AND SUCCESS** = personal values, team agreement.

This list is our translation of the ingredients contained in the Katzenbach and Smith definition of team and how those ingredients correlate to traditional documents found in a product development environment. We hope that this correlation helps you connect the definition with the documentation you may already have.

Team Building and Teamwork

The Three-Arm Framework for Successful Teams

Support from Management

- Appropriate training and skills development.
- Rewards and recognition that are consistent with the objectives.
- Feedback and guidance to help the team succeed.
- The right tools, equipment and budget to enable the team to achieve its objectives.
- Open and clear communication.

Successful Teams

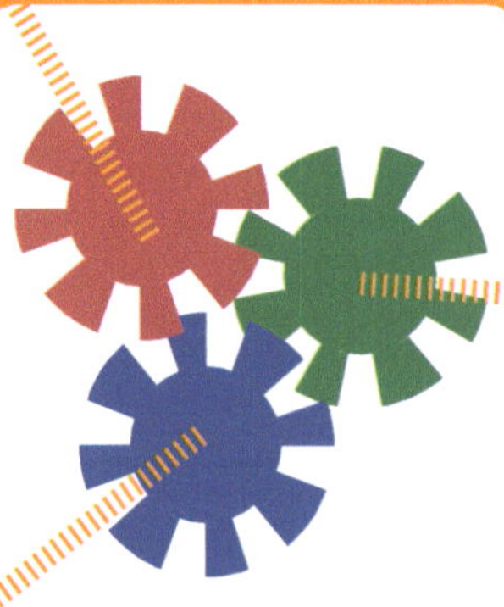

Support System

- Made up of processes, skills and tools.
- Facilitation that enables the team to succeed.
- Problem solving skills.
- Communication skills.

Critical Ingredients for the Team

- A clear set of objectives and deliverables.
- A common purpose fully supported by all team members.
- The right number of team members, with the right set of skills.
- Clear roles and responsibilities.
- Team norms, protocols and guidelines.
- The right progress and outcome measures to track progress and results.
- Open and clear communication.

Teams need a favor environment to succeed — a framework made up of support from management, the team's expectations of itself, and a support structure made up of skills, processes and tools. The above lists, though not comprehensive, exemplify the kinds of support needed. You and your organization should develop a system that is appropriate to your culture and the business objectives of your organization. If your company doesn't yet have a support mechanism for teams, we suggest that one be created in haste!

The Core Team and Extended Team

The concept of a Core Team, sometimes called a Steering Committee, emerged as part of another trend called Concurrent Engineering or, as some call it, Simultaneous Engineering. The development of new products or the management of complex projects requires the participation of various functional groups as well as contractors, consultants, suppliers, etc. The communication and coordination challenges presented by these complex situations demanded new approaches to managing projects and product development.

The formation of a Core Team empowers a small group of team members to serve as the focal point for all activities, including:

- Decision-making within clear boundaries established at the onset of the project.
- Responding to customers' expectations.
- Defining team values and team development.
- Communicating vertically and horizontally.
- Ensuring parallel work — i.e., concurrent deliverables as allowed by the project objectives.
- Systemic thinking — i.e., the big picture.
- Defining the system of metrics that track progress and outcomes.
- Owning and improving the process — for example, the product development process.

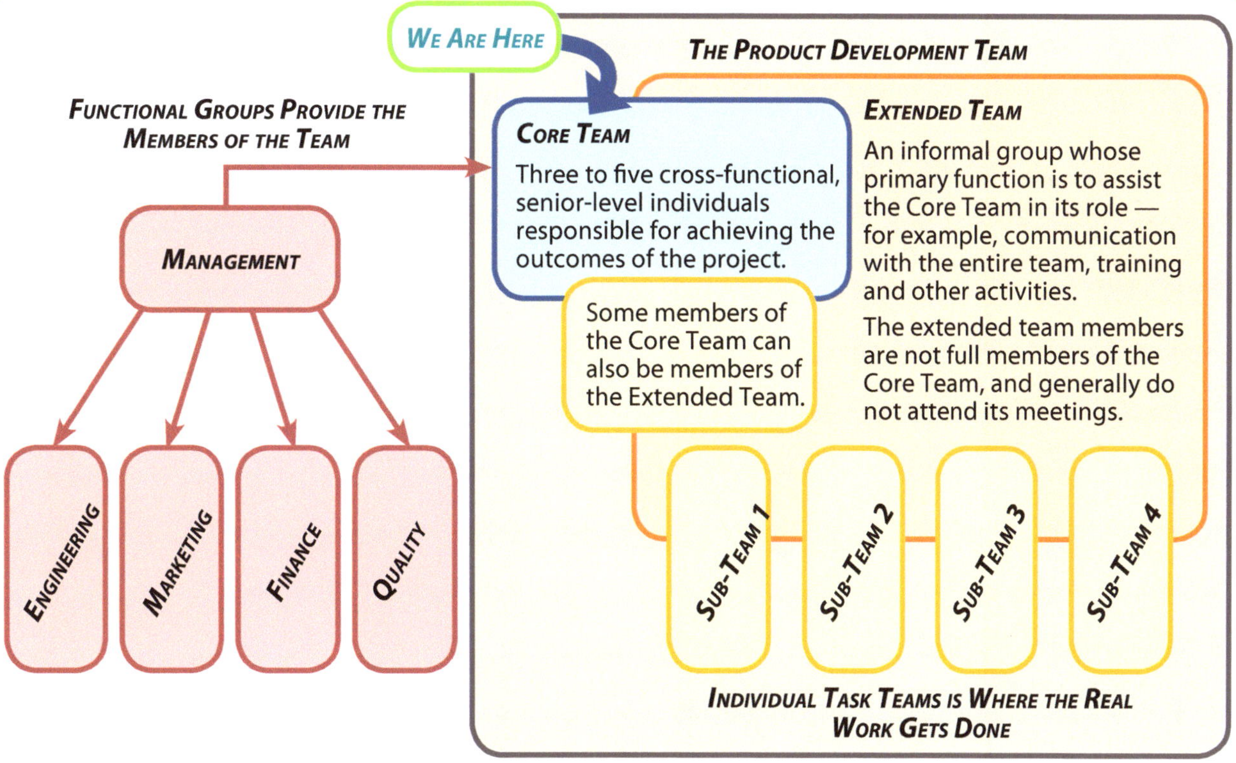

The above diagram shows the framework for using the Core Team and Extended Team concepts. The value of the Core Team is its small group size — better able to set the direction and manage the day-to-day operation of your project. Note that the Product Development Team is the largest, and that the Extended Team is a subset of the Product Development Team.

Team Building and Teamwork

Membership in the Core Team and Extended Team

Membership is the Core Team

The Core Team is generally made up of cross-functional, senior-level individuals who together represent the entire project. That is, the Core Team involves all the major functions and entities that will collaborate in bringing the project from start to results.

Membership is dictated by the complexity of the project and the level of complexity in the organization. Because the Core Team is the entity that is chartered with collaborating to achieve the objectives of the project, it makes sense to include the key representatives of the major functions involved in the project. In product development; for example, the Core Team might be made up of one representative each from engineering, marketing, manufacturing and finance, with a project manager serving as the chair of the team.

In an IT environment, the Core Team might include the technical lead, the lead analyst, representatives from the major internal customers and, of course, the project manager as the team's chair.

Key elements of the Core Team include:

- Selected by the management team.
- Reports to the management team.
- Must have one chairperson — i.e., the one person who leads and provides overall coordination.
- Must have a full scope of the project.

Membership in the Extended Team

Due to the complexity of large projects, even the Core Team may not suffice; enter the Extended Team, which is an informal group whose function is to assist with communication with all the stakeholders in the project. The Extended Team may also have other assignments as requested by the Core Team. Together, the Core Team and the Extended team provide a solid framework for communication and leadership.

Typical Process for Identifying Team Members

Selecting team members is not a trivial activity considering how busy people are. For team members to be useful to your team they must have the time available to participate, the skills to add value and the support of their managers.

Examples of Team Responsibilities

Examples of Core Team Responsibilities

- Setting up the Core Team interface.
- Setting up the team plan: work statement, schedule and engineering management plan.
- Developing the overall process and phase gates.
- Producing project decomposition.
- Selecting task teams and reviewing their progress.
- Setting up business specification (contract) and task specification with Task teams.
- Monitoring the product review.
- Responding to the task team's lifecycle.
- Assisting with Task team integration, planning and scheduling.
- Monitoring the Task team and conducting task completion reviews.

- Fostering team enthusiasm.
- Maintaining the focus on the final outcome; i.e., results.
- Ensuring training and development as needed by the project.
- Procuring and deploying tools and methods.
- Maintaining the project schedule.
- Responsible for the bounding box.
- Managing the project budget.

The above list shows examples of the responsibilities of the Core Team. More or fewer responsibilities are determined by management in collaboration with the Core Team.

The Extended Team Responsibilities

- Actively participating in the project by assuming specific tasks and sub-projects.
- Serving as the communications arm to the rest of the organization.
- Serving as rumor control for the rest of the organization.
- Actively creating a positive perception about the project within the rest of the organization.
- Participating in the meetings of the Core Team when invited.
- Lead the rest of the organization in the achievement of the scope of the project.

Team Building and Teamwork

Detailed Process: Identify Team Members

WHERE YOU'RE AT

At this point you have created the project charter and are ready to identify the members of your team.

Important

Selecting members for teams is not a trivial task. Team members should have a set of skills and level of experience that will enable them to contribute. But people are very busy with their current assignments, and they may not have the time to add yet another responsibility.

Many teams have failed because they selected the wrong team members. Avoid this situation by taking the selection seriously. Ensure that your team members have all the attributes needed to help you deliver measurable results. Work closely with managers and supervisors.

1

TALK TO THE FUNCTIONAL MANAGERS

Functional managers and supervisors are asked to nominate members based on the mission or goals of the project; thus, it is assumed that all managers and supervisors have been trained on this task.

This is an area of risk; for example, managers might not take the matter seriously enough or they might over-book the team member. Ensure that there is clarity between you and the manager.

Criteria that managers should use to propose or nominate team members

1. Does the person have the abilities? For example, the ability to effectively collaborate with other team members? Or even the availability and time?
2. Does the person have the skills? For example, computer skills?
3. Does the person have the knowledge? For example, a background in finance? Or the knowledge to understand the steps in a complex process?

2

WORK WITH THE FUNCTIONAL MANAGERS

Collaboration is needed between the functional managers and the project manager to narrow the selection and ensure balance in the overall operation of the company or organization.

Some trade-offs may be needed to meet the current assignments of the employees.

3

CLOSE THE LOOP

The project manager follows up with written communication outlining the role and responsibilities assigned to the team member.

Include a summary of the agreement or understanding between you and the manager, as well as the agreement with other managers who have supplied team members.

Team Building and Teamwork

Group Dynamics: From a Group to a Team to Results

Tuckman's Model and Product Development

Dr. Bruce Tuckman, a respected educational psychologist, first described the four stages of group development in 1965, soon after leaving Princeton: Forming, Storming, Norming, Performing. Looking at the behavior of small groups in a variety of environments, he recognized the distinct phases they go through and suggested they need to experience all four stages before they achieve maximum effectiveness. He refined and developed the model in 1977 (in conjunction with Mary Ann Jensen) with the addition of a fifth stage: Adjourning. While others have attempted to adapt and extend the model, we recommend you stick with these classic five stages.

Tuckman's model shows the progression that every team must go through to deliver results; it is your responsibility to understand these phases so you can adapt your leadership style. Don't be frustrated if you see confusion during the storming phase. Rather, support your team and allow them the time to get pass storming.

Tuckman's model is particularly important when your team is scattered. It takes more time to form a team when members are located in multiple sites. Language, culture and time zones create additional separation.

We recommend a face-to-face meeting in order to launch a team. This is particularly true when you are leading a multi-ethnic or a multi-site team. A face-to-face meeting will add velocity to your team. It will improve communication. Most of all, it will create a common purpose. While it may seem expensive to bring a team to a single location, the cost of not doing it is even higher; it shows up in delays, friction and much aggravation.

The Five Stages of Tuckman's Model

1. **FORMING:** Individual behavior is driven by the desire to be accepted by others and avoid controversy or conflict. Serious issues and feelings are avoided. This phase is characterized by team organizational issues such as when to meet, how to communicate, documentation, etc. At this point, individuals are also gathering information and impressions about each other and about the scope of the task. This is a comfortable stage, but not much progress is achieved.
2. **STORMING:** Individuals in the team can only remain nice to each other for so long, particularly as important issues start to be addressed. As patience breaks, minor confrontations arise. These may be related to the work of the group itself, roles and responsibilities. Some will want to stay in the comfortable Forming stage. The need for structural clarity sets in; people search for a framework.
3. **NORMING:** The scopes of the group's tasks are clear and agreed. Having had their arguments, they now understand each other better. They start to listen to each other better. And they begin to appreciate and support each other.
4. **PERFORMING:** Interdependence becomes a characteristic of this stage. People know each other well enough to be able to work together. They have adjusted to each other enough to allow independent activity. The group now has an identity and loyalty to the team and each other are present.
5. **ADJOURNING:** Completion is another name to this final phase. The group needs to recognize and celebrate their achievement before moving to the next assignment. The need for closure is real, as is the need to recognize accomplishment.

Questions to Ask

- **FORMING:** What is the task? Who is going to do it? Who will learn about the other team members?
- **STORMING:** What are the conflicts about? Who is in conflict?
- **NORMING:** What is the acceptable behavior? What method will we use? What is our common purpose?
- **PERFORMING:** How can I help? How can we collaborate? What else needs to be done?
- **ADJOURNING:** When is the celebration? Who needs to be recognized? How will we achieve closure?

Speak Accomplishments: How to Keep Your Team Energized

The Importance of Positive Recognition

Celebrating accomplishments is an appropriate way to energize your team. It's not often that we stop to look at the progress we have made. Worse yet, we seem to avoid the opportunity to speak of achievement, instead talking about the errors, missed opportunities and failures.

Generally, there is a misunderstanding about the need to celebrate progress. In some cases we think it's inappropriate. Nothing could be further from the truth. Teams appreciate the opportunity to feel good about hard work. It is not boasting, but rather recognition of fact!

Speak Accomplishment is the formal name of an effective tool for keeping your development team energized. Using Speak Accomplishments is a way to enable the team to express pride in craftsmanship. It is the well-deserved celebration of achievement due to dedication and hard work. It is also a way to counter any feeling of insufficiency or inadequacy that surface during the life of your project.

Advantages of Speak Accomplishments

- It energizes the team.
- It helps overcome any gloom and doom feeling during the tough times.
- It allows for well-deserved opportunities to feel proud of progress and achievements.
- It brings to everyone's attention achievements that may have been overlooked.

General Rules for Speak Accomplishments

6. Speak Accomplishments must be deliberate. It will not happen by itself. You need to be proactive and make it happen
7. You should Speak Accomplishments several times during the life of your project, especially when the team feel at a loss or gloom sets in.
8. Speak Accomplishment is not boasting. It is a rightful activity that will be welcomed by everyone in the team, even by those that are reluctant to participate.
9. It is not complicated to make Speak Accomplishments happen. It only takes 10 to 20 minutes — a small price to pay to energize your team.

Dealing with Cultural Differences

The Issue

Speaking about one's accomplishments is not culturally acceptable in many countries, and even companies. For various reasons, it has not been cultivated or encouraged. Thus, many people will be shy about what they perceive as boasting or bragging, which is not acceptable to them.

What to Do

First, legalize it by reassuring everyone that it is OK to talk about accomplishments. Explain that an accomplishment is not the same as boasting; an accomplishment is simply pride in your craft and, furthermore, it is factual and not made up. You need to give your team members license to talk about accomplishments.

Second, use Yellow Stickies. This will enable those from foreign cultures to overcome their reservations. It may also be helpful if you collect all the yellow stickies and then read them out loud. You might post the yellow stickies on a flip chart or white board for everyone to read and celebrate. Focus on the team.

Dealing with Personal Shyness

The Issue

Many members will speak freely about the accomplishments of the team, which is acceptable and very commendable, but are shy about speaking about their personal accomplishments.

What to Do

Ensure that you request two types of input — team accomplishments and personal accomplishments. Write them on separate flip charts. You may need to encourage team members to speak about their personal accomplishment; it does not come easily to many people.

Team Building and Teamwork

Detailed Process: Speak Accomplishments

1

Decide to Do It

Speak Accomplishments are not difficult, nor do they take much time.

You can do it almost any time; for example, at the start of any meeting — regardless of the meeting's topic.

You do not need permission. Take initiative, and do it! Your team will thank you for it.

Ensure that the entire team is present. Take advantage of meetings where all team members are present.

2

Set the Stage at the Start of the Meeting

Tell the team that it is important to review accomplishments.

Some team members may be reluctant to speak accomplishments because they see it as bragging. Reassure the team that it is OK to talk about the results of the hard work and craftsmanship they have shown.

Consider the ethnic and social make up of your team. In some cultures, public recognition is not appreciated. Do your homework to avoid embarrassments or a negative impact on a well-intentioned effort.

3

Provide Instructions

Ask each team member to take two yellow stickies. Instruct them to write on one of the yellow stickies an accomplishment by the team that they feel good about. On the second yellow sticky, instruct them to write a personal accomplishment.

4

Collect the Yellow Stickies: Round 1

Ask each team member to first read the yellow sticky about **THE TEAM'S ACCOMPLISHMENT** and then place it on the flip chart or the whiteboard.

5

Collect the Yellow Stickies: Round 2

Ask each team member to first read the yellow sticky about the **HIS OR HER PERSONAL ACCOMPLISHMENT** and then place it on the flip chart or the whiteboard.

6

Review and Celebrate as a Team

Read some of the yellow stickies out loud. Be sure to read both personal and team accomplishments.

Allow the team to comment and celebrate. Encourage cheers, applause and other expressions of celebration.

Encourage additional yellow stickies. Allow team members to add more accomplishments.

Continued

Team Building and Teamwork

Where You're At

At this point you have two sets of yellow stickies — one set of team accomplishments and another of individual accomplishments.

You and your team can now prioritize these yellow stickies so that you can document, communicate and archive your progress.

Cont'd

Important It is always a good idea to prioritize all the accomplishments and then share them with managers, supervisors and executives. They will appreciate the hard work you and your team are doing.

7 Communicate to Management

Work with you team to select a small number of yellow stickies that best represent the accomplishment of the team.

Develop a presentation or report.

Share the results with your team sponsor, management and supervisors.

8 Archive and Communicate

Archive the information for later use; for example, when doing a debrief of the project or when creating your final report on the project.

Learn and Improve (Kaizen)

After the first time, you will feel comfortable and will want to repeat the process during future meetings.

This is a fine way to do team building because it is the team who develops the materials, and it is the team who celebrates.

Four Phases of Flexible Project

The Typical Development Program

Proactive Project Management

Before you begin, we need you to fully understand the principles and methods of proactive product management. The role of the project manager is to anticipate and prevent anything that will limit the achievement of the project objectives. Thus, it is critical for the project manager to proactively seek, investigate and otherwise spend considerable energies being an activist — that is, the person who purposely asks questions, calls meetings and challenges the assumption that no news is good news.

There are two parts to proactive project management: the philosophy and the methods.

The Philosophy

To be proactive means that your approach is a personal commitment to ensure that progress is made — as opposed to a passive, sit-back attitude. You must be willing to make a splash, rock the boat, upset the apple cart... Enough clichés already! In short, it's all about your demeanor and the way you create urgency within the organization. It's about your ability to persuade everyone to move forward.

What to Do

Being proactive includes the old-fashioned Managing by Walking Around (MWA) and talking to the project team, stakeholders and customers about the current situation and future state. In addition, your project status meetings, presentations to management, reviews and other gatherings are also prime opportunities for exercising your initiative. Call meetings, talk to people and act with a sense of urgency, but do not sit back and wait! At the end of the day, it all rests with you and your initiative.

The Methods

After the Planning

Once all the planning activities have been performed, it is the responsibility of the project manager to take the initiative to keep his or her finger on the pulse of the project.

Once project **PLANNING** is completed, it's time to **DO** (execute), **CHECK** (review, measure) and finally **ACT** (improve, change).

The Plan-Do-Check-Act (PDCA) Cycle

Dr. Edward Deming's Plan-Do-Check-Act cycle is very appropriate and in full harmony with project management. Dr. Deming saw it as a never-ending, closed-loop way to improvement.

Definition

- **PLAN:** Your original (initial) plan needs to be constantly reviewed and updated. That's a given in product development. The objectives stay constant and drive the types of changes you must make.
- **DO:** After planning and re-planning you proceed to implementation.
- **CHECK:** You must measure your progress towards your objectives. Checking is all about tracking movement towards your objectives.
- **ACT:** Based on the measured progress toward your objectives, you must act to review your plan. And then start the cycle over again.

In the turbulent world of product development, the plan will necessarily change. Your focus should always be on the objectives of your development, which stay fixed until your executives decide to change them based on your recommendations. Changing, adapting and flexing are not bad things; rather, they show your leadership and focus on objectives.

The Typical Development Program

Five Questions of the Proactive Cycle of Project Control

With any project, the challenge is to stay within the scope or adapt to deliver your project's intended results. You can implement the proactive cycle of project control by constantly asking and answering five key questions:

1. What is the variance between the baseline and the actual (schedule, budget, etc.)?
   - The purpose of the first question is to establish the variance — that is, whether the project or a specific activity is ahead of or behind the predicted status. This can be applied to schedule, budget, resources and the overall scope of your project.
2. What are the issues causing the variance?
   - The second question seeks understanding of the issues causing the variance — that is, the root cause of the problem.
3. What is the current trend or preliminary predicted status if we continue with our current performance?
   - Given current variance, productivity and issues, the third question strives to identify where the project may end up in the future. How does the variance impact the results or outcome of your project?
4. What new risks may pop up in the future, and how could they change the preliminary predicted status?
   - However, since the future is unknown, it is better to explore now what new risks may pop up in the future and further endanger your project's delivery or results. This is the focus of question four.
5. What actions should we take to prevent the predicted state from happening and deliver on the baseline?
   - Finally, question five nails down the most important step: take action! Question five specifically probes for how to act in order to eliminate any variances, disturbances and derailments that could interfere with your project's delivery or results.

Overview of the Four Phases

To ease the navigation and usability of this guidebook, we have divided the journey of managing a project to develop a new product into four separate phases:

1. **DEFINE** the project.
2. **PLAN** the project.
3. **IMPLEMENT** the project.
4. **CLOSE OUT** the project.

Our intent is to help you understand and manage a product development project by describing how things happen in a real environment.

It is important to note a significant distinction. Two things happen at the same time: you are managing a new project and you are using the project to develop a new product. The development of the new product is only a part of delivering a new product to the market. Generally, your role as a project manager is to ensure that all the elements of a new product introduction are ready when the new product is also ready. This require you to ensure collaboration among several functions, which include marketing, manufacturing, QC, engineering, sales, finance and several others.

The area of responsibility varies from company to company Some companies wish to have you concerned only with the development of the new product, and the rest is assigned to other project managers or to the functional managers of each organization. Ensure that you know, with clarity, what your responsibilities are to avoid misunderstandings.

The Typical Development Program

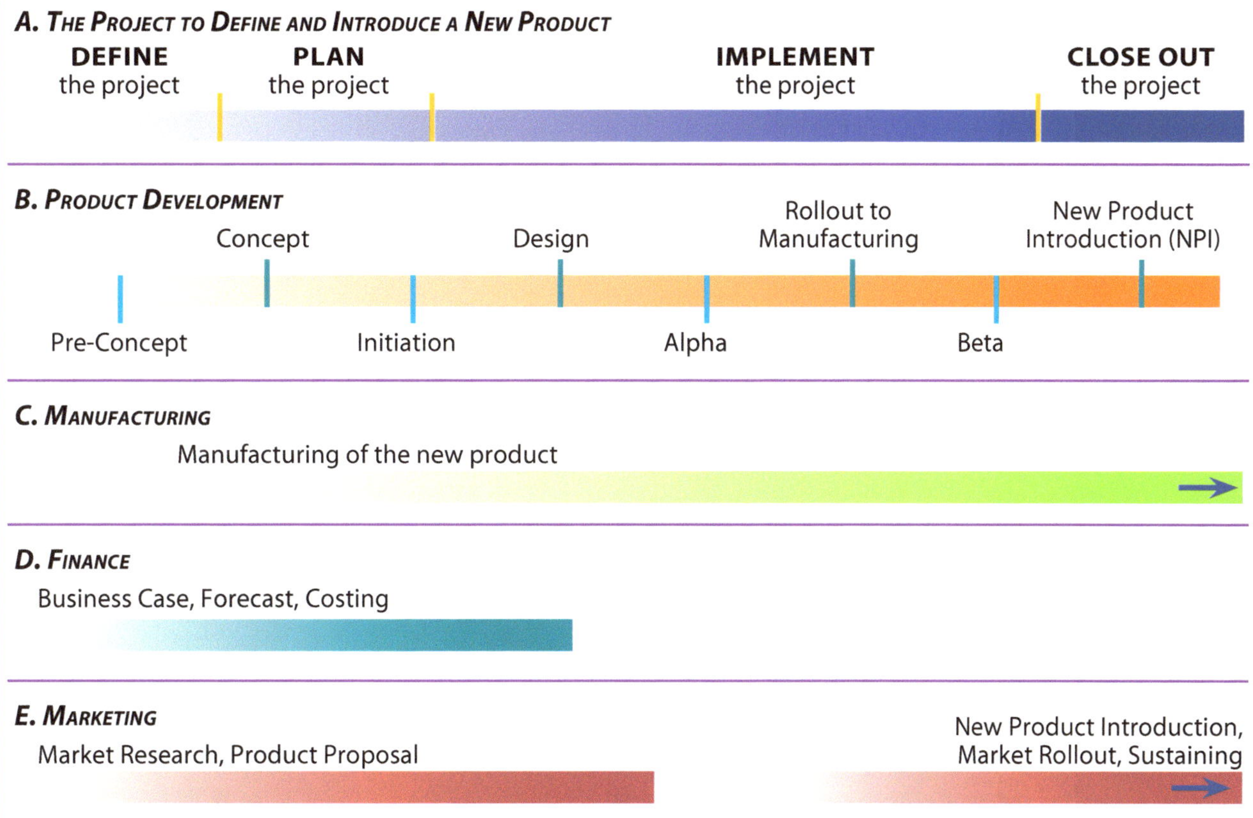

The above conceptual diagram illustrates the important difference between your project (A) and the development of the new product (B). This difference is important because it shows areas of responsibility. As a project manager, you are responsible for all the activities that will be needed to get the new product to market and deliver a profit. Note that other functions in the company (C, D and E) are involved in the project. The four phases of managing your project is the way we have chosen to organize this guidebook.

Phase 1: Define the Project

In this initial phase, you research, document and validate the project you are about to start. And in product development, it is the phase, before design, when the characteristics of the new product are being documented.

In product development, the project definition starts as soon as the product has been approved for development. In larger organizations, new products are approved in quarterly meetings tied to the Product Roadmap or similar reviews. It is a common mistake not to start the project definition as soon as the product has been approved for development. Be certain that you have a way to initiate your project early enough in the product definition so as not to miss important milestones.

Developing a new product is a cross-functional task involving Marketing, Engineering, Manufacturing, Finance — not to mention contractors, consultants and other outside resources. Defining the project to develop a new product involves all the resources that will be utilized during the development and introduction of the new product; consequently, your project must include all aspects needed to define, develop and take the product to market.

The Typical Development Program

Phase 2: Plan the Project

Once the definition of your project has taken place, the next step is to articulate all the activities that must happen in order to complete your project and deliver results.

This is the phase when your project plan takes place. It starts when the product has been approved for development and the marketing organization is defining the new product. While marketing is developing the MRD (Marketing Requirements Document) or similar documentation, you are developing the project plan, which includes schedules, budgets, responsibilities, etc. Our suggestion is to start your project planning at the earliest possible time, but avoid products that have not been approved for development.

Developing a project plan for the development of a new product is challenging due to the complexity of product development, flexibility is the secret. The ability to plan, but also to be flexible as product development is subject to many variables, many of which are beyond your control.

Phase 3: Implement the Project

With all the planning completed, it's time to implement your plan and start working towards completion and measurable results.

In tumultuous environments, such as product development the first order of business is to understand the full meaning of Proactive Project Management, maintaining the focus on the end result while managing the day-to-day challenges associated with a new product.

During execution, communication through various forms of status reports is critical — particularly if the product is being developed in several sites around the world. Also very important is your flexibility and adaptability in order to deal with changing circumstances, while maintaining your commitment to delivering a new product on time.

Phase 4: Close Out the Project

The time to finish your project is at hand. It is time to measure results and learn from the experience.

Before your team moves on to yet another development project, you must capture the lessons learned in order to benefit future projects. Product development is a fast-moving environment, and the tendency to move quickly to the next project is ever-present. You must prevail by including closure as part of the project, from the very start.

Project debriefs (postmortems) have been shown to be the best way to improve future performance; it is a shame to miss the lessons and experiences of a project.

Phase 1: Define the Project

NEW TOPIC

Project Definition

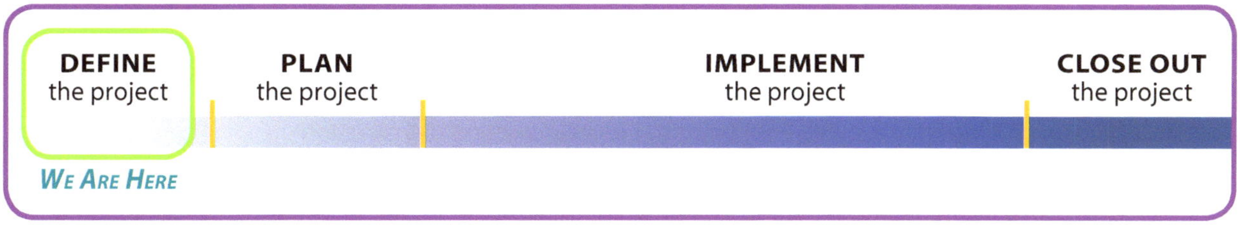

The definition phase is probably the most chaotic. You are not only defining the new product, but also the project. The critical distinction between *product* and *project* tends to confuse even the best of project mangers, complicated by the fact that the two terms are simultaneously in use. You must keep this in mind and set about leading both. While the definition phase is generally the domain of the marketing organization, it is also part of your project.

Every project has a fuzzy front end — that phase where there is much confusion because there is so little information. People are asking you many questions for which there are not yet answers; this tends to create turbulence in your team. **IT'S OK TO HAVE A FUZZY FRONT END PHASE.** Every project has it, and yours is no exception.

During the definition phase, your role is primarily to obtain answers to the myriad of questions; for example:

- What's the budget?
- Who are the members of the team?
- Is this project real?
- When do we launch the design?
- When do we launch the product?
- What is the product definition?
- How many engineers will we need?
- Can we hire contractors?
- Are other sites involved?

Essential: "Invest time early in your project. This gives you the most leverage."

The above list can go on interminably, yet you have few answers. The best thing to do is to call a meeting, inviting everyone whom you believe to be part of this development project. The single purpose of the meeting is to accumulate questions. This will provide some feedback to your team, allow them to understand that the answers will be forthcoming and start the process of team building.

These Q-minus-A sessions are very simple. Use yellow stickies and ask everyone to write down any and all questions they have about the project. Ask everyone to read aloud his or her yellow stickies so that everyone else can hear. Post each yellow sticky on a flip chart or whiteboard.

You can take one additional step; create an affinity diagram to help you organize the information. Remember that you may end up with several hundred yellow stickies. When you are finished with the affinity diagram you will have several major categories, with many questions listed under each.

You now have a very good idea of the type and amount of information you need to obtain such that you can start the project. This is a very good way to shorten the fuzzy front end phase.

See Also: Critical Tools: Affinity Diagrams

Phase 1: Define the Project

Start with the Right Questions

Initial Question: What are We Being Asked to Do?

The obvious question in every development project is, "What are we being asked to do?" Simple, yet daunting as it encompasses every aspect of your project — every deliverable — even those that cannot be articulated. The role of the project manager is to lead the organization in answering this question.

Seven Standard Questions

Wisely, the authors of modern project management broke down our initial question into seven areas to ensure that we clearly define our project. These seven questions represent an inescapable list that every project manager must create in order to lead the team to success.

Traditional Project Management Questions
1. What are you being asked to do?
2. Who is going to do it?
3. When do we need it done by? By what date? What is our timeline?
4. How much will it cost?
5. How will we manage changes?
6. What can go wrong? And what are we going to do about it?
7. How do we keep everyone informed?

These are the seven critical questions you must answer in order to lead your development program. These seven questions are from the standard project management practices.

Phase 1: Define the Project

But Wait, There's More!

For product development, we have added two additional questions that are generally not mentioned in other texts. We added these two questions because we believe that in product development there must be particular focus on the customer and the need to deliver profits to your company.

Additional Questions	
8. What is the voice of the customer?	The voice of the customer (VOC) is the list of requirements that if met will result in a new product that produces profits for your company. Without the voice of the customer there is no reason to develop the product. And while it may not be fully known at the start of your project, it must be clearly defined for your project to have any meaning.
9. What are the expected business results or outcomes?	The reality is that every development program is approved because it delivers specific outcomes for your company; i.e., profits. The focus of your development project is not on completion, but on results. Completion means that you will have a product by some specific date. Results or outcomes means that the product will deliver profits for your company, which happens weeks or months after completion. This is a critical distinction that wise project managers always keep in mind, knowing it will enable you to lead much more effectively.

Now we have a complete set of questions — these nine.

The Toolset for Project Definition

For each of the above questions there is a set of tools, templates and processes. This is a well documented and mature toolset. Many books and handbooks have been written on the subject. We recommend Dragan Milosevic's book, The Project Management Toolbox as an excellent reference.

Critical Questions	Tools for Answering the Questions
1. Who is going to do it?	Responsibility matrix.
2. When do we need it done by? By what date? What is our timeline?	Gantt chart (scheduling).
3. How much will it cost?	Cost planning and budgeting.
4. Change happens! How will we stay sane?	Change order management.
5. What can go wrong? And what are we going to do about it?	Risk management.
6. How do we keep everyone informed?	Status reports and communication.
7. What are we suppose to do?	Work Breakdown Structure (WBS).
8. What is the voice of the customer?	Customer focus and quality management.
9. What are the expected business results?	Documented business objectives of the new product.

Phase 1: Define the Project

Typical Progression for the Development of the Project Scope

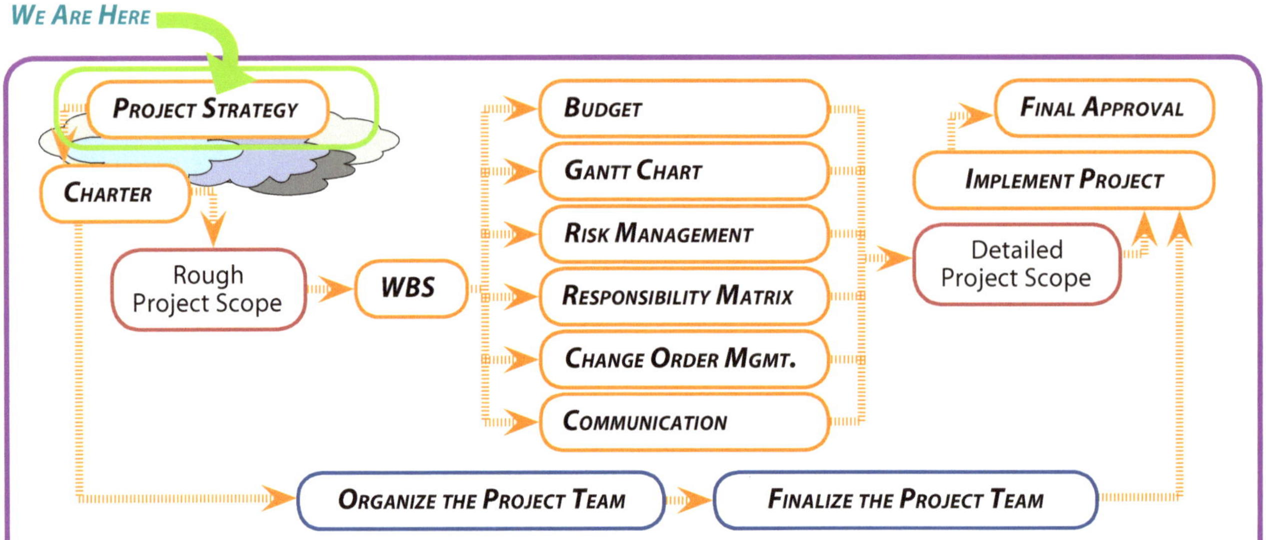

The above diagram represents the general journey to clearly define the scope of your project. While the above sequence is not universal, it is the generally accepted path. You must use your judgment to create the best path for your team given your business objectives and the culture of your team.

When you start at the fuzzy front end, the information is minimal. As you proceed, clarity begins to set in until you have a detailed scope.

In product development organization, the detailed scope varies. Many documents that are required. The important point is that you, as the leader, must have all the information you need to lead to success; i.e., the delivery of the business results of your new product

The Detailed Project Scope

The detailed project scope is *all* the information, documents and protocols that as a whole enable you to lead your project to success. To stress a point, the scope is more than a single document or even a set of documents — it's all the information, the cumulative impact of which is to enable you and your team to deliver results.

> **Essential**
> "The detailed scope for your development project includes *all* the documents that you and your team need to deliver measurable results."

Key Concepts

1. **CRITICAL SUCCESS FACTORS (CSFS):** Those few things that must happen successfully in order to deliver the expected business results. CSFs are generally five or fewer and must be essential to the success of your project.
2. **THE BUSINESS OBJECTIVES:** The measurable outcomes that summarize the anticipated business results. They can be high-level, but they must clearly articulate the anticipated outcomes — i.e., results.
3. **THE PROJECT MISSION:** A short but clear statement about the reason for your project to exist. Some refer to it as the purpose of your project. It provides the "why" of the project — it tells your team why they are working on this project.
4. **THE IRON TRIANGLE:** The project budget (cost), the timeline (schedule) and the project quality requirements. Quality is defined as the project objectives.
5. **MAJOR DELIVERABLES:** Simply, levels 0, 1 and perhaps 2 of your WBSs. The major "whats" or deliverables that make up your project.

6. **PROJECT MILESTONES:** The key events of the project against the timeline. A milestone is defined as a point in time or an event, and its importance lies in it being a peak point or the culmination for many converging dependencies.
7. **KEY ASSUMPTIONS:** A set of facts or beliefs that may impact your project in a negative or positive way.
8. **CONSTRAINTS:** A set of limits or boundaries that are nonnegotiable.
9. **EXCLUDED:** Those things your project will *not* deliver. Declare what will be excluded.

Documenting the Project Scope of Product Development

The list of documents needed for product development provides even more information. As you progress, more documents are added. Regrettably there is not standard sequence as companies vary when they require the documentation. Three common documents are the Product Proposal, the Market Requirements Document (MRD) and the Charter Document.

The Product Proposal is generally written before the project is approved for development. It has the business and marketing reasons that justify the development of the new product. This document is available to you from your marketing organization; generally the Product Manager is the document owner.

The MRD is another early document. It is the responsibility of your marketing organization. It clearly outlines the requirements, customer needs, and other customer information. This is the document that your engineering organization must have to proceed.

We suggest that you ensure that there is a Charter Document at the earliest possible time. We will study this document later in this guidebook. The Charter Document is, in our opinion, the first indication that your development project is approved and, thus, bona fide.

Use Your Judgement

Your company may have more documents that are specific to the initiation of a development project. There are many documents that can help you lead the development project. Later in this guidebook we will study most of them in some detail and highlight their role in helping you manage the project. Nevertheless, you and your team must exercise judgment and determine the set of documents that you must have in order to deliver measurable results.

On the one hand, we believe that the traditional set of documentation from standard project management is not sufficient; on the other hand we do not wish to overwhelm you with too much documentation. This is where your judgment comes into play. Ask yourself this question: "Do I have all the information I need to lead this project to measurable results?" If the answer is no, then continue to add documentation. Your team will also be helpful in determining the type and amount of documentation.

Phase 1: Define the Project

The Voice of the Customer (VOC)

Dr. Joseph Juran, the noted expert of Quality, wrote, "everything starts and ends with the customer." This is an absolute truth in product development. As a project manager, your duty is to ensure that customer requirements are clearly documented and approved and the customer is represented in every meeting and every document in your development program. The VOC is too important to leave to the marketing people; i.e., the Product Manager. It is true that the Product Manager is responsible and accountable for obtaining customer requirements. It is also true that engineers make many decisions that impact the customers. Consequently, everyone in your team must be fully aware of customer requirements.

Lack of clarity of customer requirements is a major source of complexity in product development, as no one really knows what problems your customer is trying to solve.

The Problem with Marketing Requirements Documents (MRDs)

In most product development organizations, marketing develops a MRD. Regrettably, some of these documents suffer from a set of maladies that, if not fixed, lead to confusion and product delay.

Common MRD Maladies

- **CUT-AND-PASTE:** The previous MRD is used as a source of information for the new MRD, essentially a cut-and-paste effort. Customer needs change and the MRD must reflect these changes. Copying from previous documents is a major blunder because there is no action item to interview customers.
- **MURKY MRD:** The document is not written in a way that enables engineering to design the new product. For example, the requirements are written in abstract form such as, "make it easy to use." Or the requirements are written with lack of clarity as to what the customer is trying to achieve; i.e., the problem statement.
- **TOSS-IT-OVER-THE-FENCE MRD:** The document is sent to engineering without any additional information or dialog. Rarely are MRDs sufficient to present the complete picture of the customer requirements. Additional information is needed such as Use Cases and Personas — not to mention video, photographs, diagrams, charts and other items that add clarity.
- **WRITTEN BY MARKETING FOR MARKETING:** The real issue is not the MRD itself, but the quality and quantity of information that engineering needs to design and innovate. The MRD is the formal documentation of the customer requirements that will enable engineering to innovate and design the new product; consequently, it must be written for the engineering organization and in the language of engineering. You must constantly ask engineering for feedback on the quality of your MRDs. Furthermore, you must constantly improve the quality of the MRD.

Everyone Needs to Get Involved

It is the responsibility of the marketing organization to bring the voice of the customer to the development team, but this is so important to the success of every new product that it should not be left to marketing alone. Everyone in the product development team must have great interest in understanding customers, what problems they are trying to solve and why. Marketing people should proactively encourage engineering and other members of the development team to have interaction with customers.

While it is not practical for the entire development team to interview customers, it is possible for many of them. Engineering leaders and managers must make an effort to interview and get to know customers. It must be part of the job description and expected as part of their role as engineering leaders.

Considerable amount of complexity arises from lack of understanding of customer requirements. It is a difficult task and one that must not be underestimated.

Phase 1: Define the Project

The 4-I Framework to Obtain and Use Customer Requirements

— Voice of Your Customer — | **— Value Creation —**

1 Investigate

The disciplined approach to gathering and documenting unfiltered customers, articulated and unarticulated needs through a variety of techniques that always include in-depth interviews and some level of observation of the customer's world.

VOICES

Output

At the end of this phase you should have captured (documented) the voice of the customer: the raw data from each of the interviews that you and your team conducted.

Documentation

This should be in the form of notes, photos, video and audiotape, plus the vivid memories that you and your interview teams captured.

2 Interpret

The process of organizing the customers' data and processing them into a set of prioritized and documented requirements that clearly express the customers' unsatisfied needs.

REQUIREMENTS

Output

At the end of this phase, you should have a clear, prioritized and approved list of Customer Requirements.

Documentation

Generally, this list is relatively short, between two and 10. In some cases it may be longer, but we encourage prioritization to reduce the number of the requirements, which capture the value expressed by your customers.

3 Innovate

The process of transforming requirements into one or a few product concepts that will address the customers' articulated and unarticulated needs.

PRODUCT CONCEPTS

Output

At the end of this phase you should have innovative solutions that address the requirements expressed by your customers.

This phase is where engineering and marketing collaborate to find the solutions and to innovate.

Documentation

The documentation should be a prioritized list of solutions, or a collection of product features that in aggregate (cumulatively) provide a solution.

4 Incorporate

The disciplined refinement of the product concepts into a single one validated by customer feedback, and its full articulation into a set of features and specifications that engineering can implement.

SOLUTIONS

Output

Naturally, the desired output of this phase is a profitable product.

Documentation

On the more practical side, the output of this phase generally is an engineering document, which takes into account the technical trade-offs, the timeline, design costs and other parameters. Engineering and marketing collaborate to ensure that the customer always wins; that is, that the product always has maximum value.

The Market Requirements Document (MRD) Process | The Product Requirements Document (PRD) Process

The above diagram depicts the four phases of the Master Process. It's important that you internalize the definitions, as that will help you lead your development program. Also, ensure that your team members understand the implications of the diagram. Note, the diagram is not the product development process, but rather the role of the Voice of the Customer within product development.

We offer a definition of each phase, the anticipated output and the generic documentation that serves as proof that the phase has been completed. You may already have names for the documentation, or your definitions may conflict with those in the diagram. If so, you need to harmonize (normalize) the two sets of definitions. Further, you may have documentation that is not shown in the diagram; don't discard it. Instead, analyze and compare and keep the best combination to meet your objectives. The key point is that each phase must have a tangible and verifiable document to demonstrate completion.

Phase 1: Define the Project

From Voice to Value

Every methodology dealing with understanding your customer tries to adhere to the four-phase process shown in the diagram on the previous page. It is critical for everyone who has customer interaction to understand that talking to customers is but one step in the process.

Naturally it all starts with customer interaction — the act of interviewing customers. But within this phase there are a number of tools and methods to ensure that it is done in the proper fashion. Casual meetings with the customers rarely yield usable information.

When developing a new product you must have a step-by-step process to ensure that your customer interaction produces the kinds of results you expect; that is, usable customer requirements. For example, to avoid unproductive interviews, use the Sample Selection tool. Another example, the Discussion Guide tool provides you with the script you use during the interview to ensure that you ask the right questions, in the right sequence.

Interpreting the message from the customer is a step often skipped by many organization. Customers have difficulty expressing their needs. So, the customers' messages must be carefully analyzed in order to refine the real set of requirements. Here as well, there is a set of tools and methods to help you.

Innovation is all about bringing new value to the customer. But one cannot innovate if one does not know what customers value. This is the role of the third phase in the VOC master process. Once you have deciphered the voice of the customers in the interpret phase, you are now equipped to start innovating — solving the problems that your customers are confronting.

Finally, you get to the phase where you integrate the innovation into the new product. This is the combination of requirements management, product definition and design. The art of integration is to translate the innovation form the previous step into valuable actual features of the new product.

Essential

As part of this series, we have developed a guidebook that deals exclusively with the voice of the customer in product development. Email the authors for additional information.

See Also

The Voice of the Customer for Product Development

Use Protocols to Improve Productivity and Reduce Stress

More and more, people are asked to work in teams, task forces or other forms of collaborative effort. The challenge is the false assumption that members of the teams know how to work together from the start. Don't assume that people understand the rules, covenants (agreements or contracts) and other parameters of the specific team they have joined.

While many companies use teams to solve problems and achieve other business objectives, this does not guarantee that people will know how to work together as they join various teams. This is where protocols come in to add clarity to the rules for each team.

For example, in a new team who defines how the team will communicate? Generally, we do not spend much time collaborating on the specifics of communication with team members and other stakeholders. Another example is the covenant that the team will use to make decisions — e.g., does everyone buy into or even understand the process. There are many other examples of false assumptions about how we work together, from the frequency of the meetings to communicating with the executives of the organization to managing the team's expenses.

Protocols Help

- They provide clarity by ensuring full understanding of how the team will operate.
- Protocols reduce the opportunity of conflict by having a point of reference that is available to all team members.
- They enable team members to participate by providing the rules and guidelines of behavior.

General Rules for Protocols

- All protocols must be documented and made available to all team members. It is also helpful to review the protocols during the life of the project.
- All protocols must be conveniently available to all team members. Incidentally, *convenient* is defined by the team members. No use having a protocol if it is kept a secret.
- It is essential that the team participates in the development of the various protocols. Collaboration in the development provides clarity and understanding. It must be a cooperative effort that ensures alignment — i.e., the willingness of team members to follow the protocol for the duration of the project.
- Beware of protocol proliferation — that is, creating too many protocols to the point of ineffectiveness. Generally, every team should have from three to five protocols, depending on the scope of the project.
- Protocols should be short and easy to remember. Avoid protocols that have long lists of attributes or criteria. Best to have three to five items or attributes.
- Do not develop the protocols all at once. The team may grow fatigued from developing too many protocols. Rather, space them out during the early meetings of the team, one at a time. You can always add more protocols.
- Persevere! You will encounter some resistance to developing protocols. Some team members may say, "But we have worked together before," or, "We all work for the same company — we already know how to work together." Every team is different, and as such must agree on the ways its members will collaborate.
- Protocols must be written with much clarity, and all key words must be defined in the protocol itself — for example, in a glossary.
- The need for protocols is acute when your team members are in different cities or when team members belong to different companies, such as suppliers, contractors and consultants.

Phase 1: Define the Project

- Protocols are project-specific. While company policies, SOPs, etc. are generic and apply to everyone in the organization, protocols are created to drive a *specific* project to success. No two projects are exactly alike.

The Meaning of Words

Clarity of language is a critical ingredient of every protocol. Most words are subject to interpretation due to the reader's background, beliefs, experiences, etc. Consequently, it is important that words in a protocol are clear and express the intended meaning.

This step is particularly important when the product is being developed in several sites where there are different languages and cultures at play. We believe there is considerable waste due to the lack of sensitivity to other cultures and languages. It is wrong to assume that just because the product being developed is technical in nature that everyone will have the same level of understanding. In our experience, lack of attention to communication — particularly the spoken word — leads to considerable waste.

Action Steps

- Create a glossary in which you define every term that could cause confusion or might be misinterpreted. This glossary becomes an integral part of your protocol. Ensure that everyone reads the glossary and that there is agreement on the single intended meaning.
- Make extensive use of cross referencing tools such as hyperlinks and footnotes to define words in your protocol. Hyperlinks and footnotes have the added benefit of being conveniently located where the reader needs them.
- Ensure that your team members are part of defining the protocols and the words used within them. Encourage your team members to ask questions until there is 100 percent clarity.

The Universal Rule About Words

The more common the word, the more subject it is to misinterpretation and, therefore, the more important to define it. Avoid taking the meaning for granted because it's a word that is very common. For example, the word *customer* might seem obvious, yet it is easily misinterpreted. Ensure that every member of your team has the same meaning for every critical word.

Detailed Process: Develop a Protocol

1

Decide Which Protocols You Need Now

- Start with the protocols you need right away.
- The need for protocols will manifest itself as your project moves along.
- You might also have a good idea based on previous experiences.
- We also recommend you do not attempt to develop all your protocols at one time; your team might get tired. Best to do it as needed.
- Start with a short list, perhaps three.
- Involve your team in deciding what protocols you will need.
- The key question: Who will work together to deliver results?

Important! Invite other stakeholders who might be impacted by your new protocol. Remember, this is a contract among all parties involved, some of whom may not be part of your team.

2

Define Your Protocols

- Meet with your team or assign the responsibility to a few members (sub-team).
- Define the protocol — e.g., Decision-Making protocol, Escalation protocol, etc. You will need the name and a short definition.
- Brainstorm with your team on the intent of the new protocol — e.g., "Why do you need it?" "What problem are we solving with the protocol?"
- Discuss current problems — i.e., what happens currently without the protocol.
- Start to discuss the terms of your new protocol.

3

Refine Your Protocols

- Once you have all the terms of your protocol, proceed to prioritize them — i.e., reduce the number to the critical few.
- Remember to phrase your protocol in a clear fashion so that everyone understands it, even those who did not attend your meeting.
- Reduce the number of terms to fewer than five — preferably three. Too many, and your team might have difficulty remembering.

Important! All protocols are project specific.

You must develop them with each team and only as you need them.

Continued

Phase 1: Define the Project

CONT'D

4 DOCUMENT AND COMMUNICATE

- Document your new protocol, and make it easily available to your team members and to all others who need to follow it.
- Do not keep your new protocol a secret. Ensure that all parties involved clearly understand it and agree to follow it.
- By the way, sending the protocol in an email may not do the job. It is worth speaking with each person or calling a meeting to ensure clarity.

Important

Avoid the proliferation of protocols — they become difficult to enforce, and your team might feel stifled or constrained.

Keep it to no more than five — preferably three.

5 ENFORCE YOUR PROTOCOL

- Your protocol is of no use if it is not followed.
- You need to place special effort in reminding your team members about the protocol. For example, you may review all your protocols during a staff or team meeting, or you may want to create a poster listing your protocols.
- Call any breach of the protocol to the attention of your team members.
- Ask your team members to help you enforce the protocols.
- Recognize and celebrate adherence to the protocols.

LEARN AND IMPROVE (KAIZEN)

Over time, you will find ways to improve the protocols and the way they are communicated and enforced.

Document improvement and archive notes so you can use them in future projects.

Disseminate successful protocols to other projects and teams.

Common Protocols

Remember, every team must decide how many and which protocols to develop. Here are some archetypal examples of common protocols to develop.

Communication Protocol

The Communication protocol describes how the team will communicate internally and externally. This should include email, status reports, budgets, presentations to executives, etc.

Key questions to ask:

- How will we communicate?
- How often will we communicate?
- What does not need to be communicated?

Decision-Making Protocol

The Decision-Making protocol explains how decisions will be made, as well as who makes the decisions and how the decisions will be communicated to the team and other stakeholders.

Key questions to ask:

- Who makes the decisions? And what type of decisions will they be authorized to they make?
- How do we know when a decision has been made?
- How will we communicate a decision when it is made?
- How will we know that the decision has been acted upon?

Meetings Protocol

This protocol tells the team the number, type and duration of meetings as well as who needs to attend. It also defines the reason for each meeting and its expected outcome.

Key questions to ask:

- What types of meetings will we have?
- Who will call these meetings?
- Who will attend the various meetings?
- How will we document and archive the information from the meetings?
- Who needs to know about our meetings? And why? How often? By which means?

Action Items Protocol

Describe how action items will be defined, archived and communicated with an Action Items protocol. It should clearly explain who will be responsible for an action item, when each action item must be completed and what needs to get done.

Key questions to ask:

- Who will capture action items?
- How will we know that we have an action item?
- How will we know that an action item has been fulfilled?

Phase 1: Define the Project

Escalation Protocol

Explain how the team will escalate various challenges confronting it. Teams face problems, barriers, obstacles and other trials that can prevent fulfilling the team's objectives. Deciding if something needs to be escalated is very important — as is determining who will do the escalation and how the resolution will be handled.

Key questions to ask:

- What needs to be escalated? And why?
- Who will do the escalating?
- To whom do we escalate? And why? Has this party agreed?

Conflict Protocol

It is important to define how to handle conflicts between team members or with key stakeholders. A Conflict protocol provides a safe environment for resolving conflict in a level-headed and civil manner; thus avoiding hurtful or mean-spirited dialog.

Key questions to ask:

- How will we know we have a conflict?
- What will be the immediate action once we declare a conflict?
- What is the process to resolve the conflict? Who will be involved? Who will not be involved?

Expenses Protocol

The Expense protocol defines how the budget for your project can be spent. It defines authority, the paper trail and accounting rules.

Key questions to ask:

- Who has the authority to spend the budget?
- What is the signature procedure? That is, how many signatures are needed, and who must sign?
- What documentation will be required to spend the budget?
- Who can make a commitment to spend money?

Summary

As you can see from the above list, there are many opportunities to create protocols to help your team become more effective. Any area of your project where there is confusion, friction or waste is a candidate for a protocol.

Phase 1: Define the Project

Example: Develop a Protocol

EMail Protocol for the Alpha Team

The Alpha project will be a virtual project distributed around the world. We will all have to work with people who are not in our proximity. Thus, we must deliver results though the rest of our team is scattered around the world.

Therefore, the Alpha team will strive to address the following challenges:

- Clarity about the actions expected after reading an email.
- All email will be easy to read. We will eliminate misspellings, missing words and grammatical errors that obscure the meaning.
- We will be clear about the action readers must take upon reading our email.

1. We will take full advantage of the subject line.
   - This will enable the reader to search our email more conveniently.
   - We will put the most obvious words in the subject line so that recipients can find the email using a subject search — for example, "Monday's planning meeting pre-work assignment." We will provide the reader with enough information for searching and sorting.
   - We will not start the email text in the subject line.
   - We will not be unnecessarily vague. For example, "About next week" is too vague. "Confirming our meeting next week" is more appropriate.
2. We will cover one single item of business.
   - We will send separate messages with individual subject headings — that is, if there are three subjects, then we will send three separate email messages. This makes it easier for readers, allowing them to understand, review and react to each subject logically and to simplify filing and archiving for later use.
3. We will read and edit before we send.
   - As a form of written communication, an email represents you professionally. Thus we will send only our best work.
   - We will never send an email without first checking the spelling. Because email is so informal, we tend to write casually, thus creating typos and other blemishes that make the message difficult to read.
   - Be mindful that some recipients may not speak English as well as you do.
4. We will make it crystal clear.
   - We will specify what actions are required of each recipient. For example, "John, please develop a database of all Midwestern customers by the end of this week." And, "I will need the database by Wed., Nov. 16, at 2 pm EST." Ambiguity here can cost our team precious time, not to mention aggravation.
   - An action item is made up of three parts: 1) do what; 2) by when; and 3) by whom. We will use this simple checklist to ensure that our email will be acted upon.

Protocols should be short, perhaps three items. It's difficult for people to remember long lists of requirements. We suggest you prioritize based on improving the clarity and effectiveness of communication. The list of items below is a set of suggestions and should not be used verbatim. Use your judgment and collaborate with your team to develop protocols that will actually be supported and used.

Phase 1: Define the Project

Typical Progression for the Development of the Project Scope

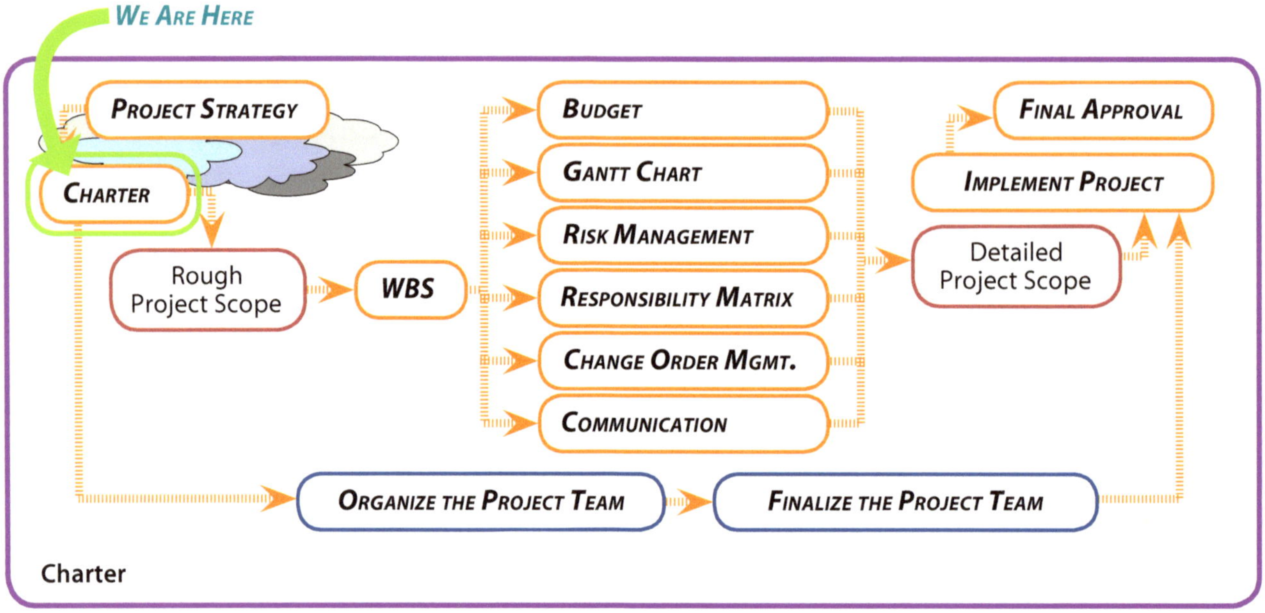

Charters

A Charter can be thought of as a combination of covenant — a description of roles and responsibilities — plus a clear outline of the desired results. It is a vehicle that takes the team from much uncertainty, to a level of commitment, and then to delivering results.

A Charter therefore, is an agreement (contract) between a team and its sponsors defining the key operating parameters and metrics of the team's performance, and most of all the desired outcome. All should be written in a language that is easy to understand by every member of the team.

Definition of a Charter

A Charter is an agreement between a team and its sponsors, defining the key operating parameters and metrics of the team's performance. Generally drafted at a Charter Formation Event, a Charter captures the rationale and project operating scope in a way that speeds alignment and decision-making in the fuzzy front end.

A Project Charter is a tool that formally authorizes a project. It defines what work you need to do in order to successfully complete the project. Charters form the baseline for further schedule and cost planning. When a Project Charter is combined with the schedule, cost planning, quality plan and risk plan, the aggregate becomes the Project Plan (Project Scope).

Benefits of Charters

- In the fuzzy front end, the conversation about Charters creates more rapid alignment between the sponsors and members of the team.
- In the early phases of team formation, Charters help the team move through phases of team maturity. It is a well established best practice to have a Charter Formation Event as an early team activity.
- During the course of the project, Charters serve as a touchstone for when the project is drifting off course or running into higher levels of business risk.

Phase 1: Define the Project

General Rules for Charters

- Charters are generally issued by a manager who is external to the project. In some cases, Charters are assigned by the PMO (Project Management Office).
- Charters must be documented.
- Charters are clear about the difference between mission, goals and deliverables.
- Charters are negotiated and agreed to in good faith. Simply put, charters are agreements.
- Charters are believed by all parties to be achievable.
- Charters are reviewed, revised and refreshed as major events occur in the evolution of the program.
- Initial Charters are between the sponsor and program manager and enable team formation.
- Subsequent Charters are between sponsors and the team and shape execution.
- The program manager facilitates the conversation.

Key Points About Charters

- Charters ensure written understanding of what you will and will not do — i.e., document the scope.
- Developing a Charters forces the dialog for clarity in understanding.
- The process of chartering is more important than the actual document.
- Remember, there are three separate aspects a Charter needs to define:
    1. What do you think is going to happen?
    2. What does your team thinks is going to happen?
    3. What does your manager thinks is going to happen?

Charters and Other Documents

Why do I need to deal with yet another document? Remember, a Charter is the first written evidence of a new product. When you start forming a Charter, very little information — not to mention documentation — exists. The only document that might precede the Charter is some form of a Product Proposal. Ultimately, a Charter coexists with other documents like the Product Proposal, the Business Plan, the Marketing Plan, the Requirement Document, the Project Scope, etc.

A charter is more about a covenant agreement among all parties such as management, engineering, marketing, program management, etc. Some call it the Up-Front Agreement. It ensures that everyone — including members of your management team — has a single point of reference and understands the boundary conditions.

Phase 1: Define the Project

Detailed Process: Obtain Approval to Use Charters

Where You're At

THESE FIRST THREE STEPS happen only once: the first time the idea of a Charter is proposed. Use these initial three steps if your organization does not yet use Charters for new projects.

If your organization already uses Charters, then jump ahead to **DETAILED PROCESS: DEVELOP A CHARTER** on the next page.

Note, your managers must understand the role of charters, and they must be willing to support them; otherwise, the process may lead to misunderstandings.

1 **Identify the Sponsor**

A sponsor is a manager or executive of your company who agrees to support your efforts by spending time with you, removing obstacles and otherwise helping to ensure your success.

Generally, the project sponsor is the executive most closely responsible for the business performance of the project, the executive who has the most to gain or lose from this project.

2 **Create Alignment with the Sponsor**

You and your sponsor meet to discuss the benefits of Charter and be in alignment.

Once in alignment, proceed to develop a plan to obtain approval for the use of Charters in projects.

Clearly identify the ways that your sponsor will help you obtain approval.

3 **Obtain Approval to Use Charters**

The purpose of this step is to obtain official approval to use a Charter document in your project.

Managers should understand the use and purpose of a Charter — not only so that they will buy into the idea, but also because there are clear responsibilities for the managers.

A **NO**

Stop and try again in the future.

B **Yes**

At this point, you have approval to use Charters in your projects.

To solidify this agreement:

- Proceed to communicate the approval to all managers and supervisors. Communicate the reasons for Charter documents. Explain to them their roles.
- Develop the templates and forms to be used.
- Introduce the Charter tool in one project to ensure that everything works properly.
- Ensure adoption of Charters across your organization.

Continue with Detailed Process: Develop a Charter.

Continued

Detailed Process: Develop a Charter

Where You're At

At this point, you have approval to use Charters in your projects.

If your organization does not already use Charters, then flip back to **DETAILED PROCESS: OBTAIN APPROVAL TO USE CHARTERS** on the previous page.

Note, your managers must understand the role of charters, and they must be willing to support them; otherwise, the process may lead to misunderstandings.

Cont'd

4 Validate Alignment with the Sponsor

Gain clear and unequivocal agreement that you will use a Charter on this project.

Gain agreement on the timetable to complete the Charter (not the project).

Ask the sponsor to communicate to his peers that a Charter will be used in this project.

Important

Collaborate closely with your future development team. Perhaps you can have a meeting just to talk about the information for the Charter.

You may not have a team yet because your project will not be approved until the Charter is official. However, find people with whom you can closely collaborate to develop your Charter.

5 Draft the Charter for the Program

Use the template Develop a Charter provided in this guidebook.

Keep in mind the complexity of the project and the amount of innovation that will be required to complete it.

Request input on the draft of the Charter from key stakeholders, your peers and other persons who can add value, critique and insights.

See Also

Develop a Charter Template

6 Have an Agreement Dialog with the Sponsor

Checklist:

- Validate the scope.
- Validate the purpose.
- Identify risks.
- Start the conversation about the proposed team members (who, when, etc).
- Ensure approval of the Charter document.

Continued

Phase 1: Define the Project

CONT'D

Important Before you have a Charter, you may have a small team whose role is to precisely define the Charter.

At first, your team may be only two or three people, but as your development project starts taking shape, your team will grow. As more team members join, the document will be available to them.

Your charter will also grow and become more useful with time.

7

FORM THE DEVELOPMENT TEAM

Key steps:

- Ensure that the managers of the team members support their involvement.
- Ensure that the team members are willing participants.
- Ensure that you have the right resources at the right time.
- Ensure that you have the budget to support the team.

Important At this point, you may not have all the members of your development team available because they may be still working on other projects.

Identify as many as possible and invite them to the kick off meeting.

8

HOLD THE KICK-OFF MEETING

The Kick-Off Meeting — explained in more detail later in this guide — is the first official meeting of the team. The only objectives of this meeting is to ensure clear understanding of the Project Charter and obtain the support of the entire team.

These meetings generally last several hours, sometimes a full day.

Examples: Charter Tool

Example 1: Abstract Charter Tool

1. Project name:

2. Short description of your project:

For example:

- ❖ The mission of your project.
- ❖ A narrative summary of the scope

3. What is the alignment of this project to the business strategy of your company?

For example:

- ❖ Why is this project important to your organization?
- ❖ How does this project link to the business objectives of your organization?

4. What are the project deliverables?

For example

- ❖ What will be the tangible end-results of your project?
- ❖ What is the visible evidence, which will tell you that the project has delivered results?

5. How will you know when the project is successful?

Be SMART:

- ❖ Specific, measurable, actionable, realistic and time bound.

6. Key assumptions about this project.

7. Key constraints; i.e., what are the boundary conditions?

Important: This list of questions illustrates the type or nature of the questions you might want to address when developing your Charter document.

The list is not a survey. It is for illustration only. You need to adapt it to your business objectives and the culture of your organization.

Roles and Authority

8. Who is the Project Manager?

9. Who is the Sponsor?

10. Who will make the key decisions?

11. Other stakeholders?

Approvals

Management Sponsor: ____________________

Project Manager: ____________________

Funder ($): ____________________

- ❖ The person or department that will pay for this project.

Definition: SPONSOR

A manager or executive of your company who agrees to support your efforts by spending time with you, removing obstacles and otherwise helping to ensure your success.

Generally, the project sponsor is the executive most closely responsible for the business performance of the project, the executive who has the most to gain or lose from this project.

Phase 1: Define the Project

Example 2: Concrete Charter Tool

This list of questions illustrates the type or nature of the questions you might want to address when developing your Charter document. The list is not a survey. It is for illustration only. You need to adapt it to your business objectives and the culture of your organization.

1. Project name:
   - Columbia

2. Short description of your project:
   - The mission of Columbia: To lead the industry by establishing a new standard of performance (25% faster than current products) and ease of use.
   - We will develop Columbia in collaboration with two suppliers in Shanghai China, the documentation will be developed in Bangalore India, and the software will be outsourced to Belarus. We will take 18 months to develop the product and make it available to purchase.

3. What is the alignment of this project to the business strategy of your company?
   - Columbia fits our corporate mission to be #1 or #2 in any of the market segments we choose to address.
   - It also provides a margin of 60%, which is fully in line with the corporate target of 50% or better.
   - This project also sets a higher standard for ease of use and will require considerable research and close collaboration between our software and hardware teams.

4. What are the project deliverables?
   - Columbia will be available for the customers to purchase 18 months from today (March 2, 20XX). It will have a margin of 60% and will be recognized by the end users as the easiest product to operate in the market place.
   - We will measure time-to-market, margin and customer satisfaction levels for ease of use and operation.

5. How will you know when the project is successful?
   - We will know that it is successful when the margin at time of introduction is 60%, the product is available to purchase by customers 18 months from today (March 2, 20XX) and the satisfaction with ease of use exceeds 80%.

6. Key assumptions about this project.
   - We will assume that at least one additional new product will be introduced by our main competitor. Also that the performance of the new competition will not match our target for Columbia.
   - We will assume that there will be close coordination and communication with all our partners in China, India and Belarus.

7. Key constraints; i.e., what are the boundary conditions?
   - We must deliver the new product in 18 months from today (March 2, 20XX)

Phase 1: Define the Project

Roles and Authority

8. Who is the Project Manager?

- Amin Hassan will lead the project.
- Lin Chao will be the project manager in Asia

9. Who is the Sponsor?

- Julio Mendoza, VP of Marketing.

10. Who will make the key decisions?

- The Core Team.
- Decisions will be escalated to the Executive Council as needed.

11. Other stakeholders?

Approvals

Management Sponsor: ____________________

Project Manager: ____________________

Funder ($): All funding and budgets will come from a corporate account.
May Lopez will manage all financial matters.

Phase 2: Plan the Project

Plan the Project

Introduction

Planning a project for developing a new product includes a documented plan, the formation of a team and extensive communication. Planning is simply answering a set of questions about your project so you can implement and deliver results.

In this guidebook, we start the planning phase with the Kick-Off Meeting; in your case this may not be the best starting point. You may want to form a small team to begin to obtain some of the information prior to the Kick-Off Meeting.

The Product Roadmap

New development projects are extracted from the Product Roadmap, which is a document generally the responsibility of the marketing organization. Marketing (Product Managers or Product Line Managers) should have enough advance notice so you can start your planning efforts. The Product Roadmap is regularly reviewed in most organizations — generally every three months — but it varies. You should attend these meetings in order to get visibility of new development projects.

The Product Roadmap will have enough information to provide you with a solid starting point.

The Product Proposal

Every new product has some form of Product Proposal — a document that contains enough information so that managers and executives can make a go-no-go decision. In some organizations, other documents precede the Product Proposal; for example, the Idea Proposal. These are very short documents, generally one page, that contain very preliminary information about proposed new products.

In summary, there is enough information in your organization to provide you with early warning of an impending development project. Become familiar with these documents and meet regularly to avoid surprises.

Kick-Off Meetings

Most, if not all, projects have a very fuzzy or murky beginning. Team members may not even know they are part of the project. Managers may not know when the project started. And confusion reigns supreme. This situation greatly contributes to wasteful activities, loss of precious cycle time and overall reduction in productivity. Kick-Off Meetings are one sure way to reduce the fuzziness!

Definition

Kick-Off Meetings go by many names: Off-Site Meeting, Retreat, In-Service Meeting, Work Session. Regardless of what it is called, a Kick-Off Meeting is the event that clearly and unequivocally announces to everyone that the project or program has been launched. It's an event that is out of the ordinary; it clearly stands out as different and separate from every other meeting.

A Kick-Off Meeting is also the forum that creates clarity about the purpose, outcome, approach, metrics, roles and responsibilities for your project or program.

What a Kick-Off Meeting is Not

1. It is not a one-way presentation. Beware of PowerPoint presentations!
2. It is not a forum to order, coerce or intimidate people to make a commitment.
3. It is not a Parade of Kings — i.e., the ritual where some executives make PowerPoint presentations that have little or no relevance to the matter at hand.
4. It is not training or a seminar.

General Rules

- A Kick-Off Meeting should allow the team to spend considerable time talking about the project. It is not a presentation of a Charter; instead, it is an open dialog that ensures understanding and commitment.
- Kick-Off Meetings must be well planned. They are not casual events — they involve many people and, more importantly, they greatly influence the results and speed of your project.
- Making a Kick-Off Meeting an event does not mean balloons and marching bands. While fun and merriment are not excluded, what makes an event is the tone you set for the meeting. For example, hold the meeting away from the customary conference room, preferably in a venue away from the office. Invite keynote speakers to talk on the subject at hand.
- For smaller projects, a Kick-Off Meeting should be at least four hours long. For larger projects, consider a full day or, in some cases, two to three days. It's a small price to pay for added velocity and exceptional results.
- Make the business case. Your management chain will ask you to justify the expenditure of time and funds — rightfully so. Link the Kick-Off Meeting to the business objectives of your project or program. For example, if your program involves large expenditures, then a Kick-Off Meeting is a small price to pay to ensure success. Another justification is that of risk. For example, what would happen if you did not have a Kick-Off Meeting? The program may not be completed on time, the team may be confused and objectives may not be clear.

Objective

First, the intent the Kick-Off Meeting is to provide your team with ample time — without disruption — to concentrate on the matter at hand; i.e., to launch your development process. Second, the intent of the meeting is to properly and officially start development activities. This is when the clock begins to tick!

Essential

"The intent of the meeting is to properly and officially start development activities. This is when the clock begins to tick!"

Phase 2: Plan the Project

These are working sessions, many times starting early in the morning and going well into the night. They are not a vacation, although some rest times may be provided. Kick-Off Meetings allow team members to relax so they can continue their work.

Stumbling Blocks

Why Do Kick-Off Meetings Fail?

There are few reason why Kick-Off Meetings fail. Do not fall into these traps:

- **LACK OF CLEAR PURPOSE FOR THE KICK-OFF MEETING.** When insufficient time and effort are spent understanding the dynamics of your group, planning the desired outcomes from the meeting and developing a comprehensive agenda, it results in valuable time spent in useless discussions during your Kick-Off Meeting.
- **LACK OF A CLEAR AGENDA.** Many Kick-Off Meetings are planned only on the most casual basis — the proverbial "back of the napkin" approach. But these meetings require considerable planning and a clear definition of roles and responsibilities. Also, these meetings require a meticulous agenda that outline every minute of the meeting. Moreover, the agenda must be realistic by allocating the proper amount of time for each item. Cramming activities is not productive, and leads to frustration. Resist the temptation to add too many agenda items. It's better to lengthen the meeting. Better still, prioritize those things that must be accomplished, and eliminate the rest.
- **LACK OF FOLLOW-UP AFTER THE MEETING.** Many well-intentioned commitments go unfulfilled. Frustration sets in when considerable effort was made by the team during the meeting and nothing happens afterward.
- **POOR USE OF FACILITATION SKILLS, OR NONE AT ALL.** Every Kick-Off Meeting must be facilitated. It is too important to leave to chance. Facilitation ensures a smooth flow and, most of all, attainment of the meeting's objectives.

Themes

Is a Theme Important?

Some organizations develop a theme for the Kick-Off Meeting; we recommend such a step. It's part of team building, and it communicates the mission of the meeting. Avoid the tired old platitudes — for example, "Defining the Future" or "Sharpening our Leading Edge." Instead, engage your team in choosing a name or theme that is relevant and inspires action and results. It's fine to be whimsical, if the theme emerged from the ranks of your team.

Management Sponsors

The Role of Management and Executives

Your project should have a Management Sponsor (Management Mentor). This individual is important to the success of every program. Therefore, it's important to invite Management Sponsor to participate in the Kick-Off Meeting. When you plan the meeting, ensure that the Management Sponsor has a visible and productive role to play.

Avoid trivial activities such as the tired old PowerPoint presentation that creates no dialog. Instead, invite your Management Sponsor to be part of a Q&A session. This will allow team members to ask relevant questions. Also, it will engage the Management Sponsor in the subject at hand. If possible, engage the Management Sponsor in some of the activities of the team — for example, developing the Team Charter or reviewing the project objectives.

Phase 2: Plan the Project

A word about the Management Sponsor. This is not a symbolic role. This is your mentor. The Management Sponsor must be committed to assisting in every way possible to ensure the success of your program, including removing obstacles, ensuring proper funding, obtaining training for your team and, in short, being committed to your success. Make certain that your prospective Management Sponsor fully understands his or her role and is committed to spend the time needed to help.

One Kick-Off Meeting for the Entire Team

What to Do with Distributed, Virtual Teams

More and more, development projects involve people who are distributed in several venues around the world.

We highly recommend that the entire team be brought together for the Kick-Off Meeting. At the very least, we strongly suggest that at least two members from each site be part of the Kick-Off Meeting. The meeting will add much clarity to all team members, build relationships and provide energy for the long run.

The cost may be high, but we encourage you to make the business case to your management chain. For example, the Kick-Off Meeting is a risk management step, and having the team meet at the start of the program reduces risk by adding clarity. You can also use the Kick-Off Meeting to identify risks that might delay your program. Team members from other locations will provide valuable input in this area.

Essential

We highly recommend that the entire team be brought together for the Kick-Off Meeting. At the very least, we strongly suggest that at least two members from each site be part of the Kick-Off Meeting.

Phase 2: Plan the Project

Detailed Process: Plan and Facilitate a Kick-Off Meeting

1

Form a Core Team for Your Kick-Off Meeting

A Core Team is a small group of people who serve as the steering committee for the Kick-Off Meeting. This team will meet before and after the meeting to ensure continuity.

Call the first meeting of the Core Team to define the objectives of the Kick-Off Meeting. Define the outcomes or deliverables from the meeting. Assign roles and responsibilities. Define the measures of success.

Coordinate logistics such as venue, meals, audio/visuals.

Important: A Kick-Off Meeting can last from half a day up to three or more full days. Generally, for a new product two days is appropriate. This is one meeting where ample time is highly recommended. This is particularly true if you are bringing participants from around the world who may not see each other again until completion.

See Also: Core Teams

2

Solicit Input from the Participants

Interview a sample of the participants; the Core Team can do this. These are informal interviews aimed at achieving the following results:

- Understand the dynamics of your group. For example, their expectations from the Kick-Off Meeting, and their concerns and reservations.
- Build anticipation for the meeting. Participants will be aware of the forthcoming Kick-Off Meeting and start looking forward to it.
- Identify barriers and obstacles — i.e., anything that can interfere with the completion and success of the Kick-Off Meeting.
- Elicit agenda items and topics for discussion.

Important: Start the process at least four to six weeks prior to the actual date of the meeting. This will allow proper planning, scheduling by the participants, selection of the venue, etc.

3

Create the Agenda

Once you have received input from the participants through interviews, start developing the agenda for your meeting in full collaboration with the Core Team.

- Communicate to all participants everything they need to do prior to the meeting.
- For key questions and logistics, see Kick-Off Meeting Planning Tool template.

See Also: Kick-Off Meeting Planning Tool

Continued

Phase 2: Plan the Project

CONT'D

4

INVITE PARTICIPANTS

Ensure that all participants have ample notice to plan to attend your meeting. Three to four weeks is suggested.

Ensure that all the participants understand their role in the Kick-Off Meeting — i.e., those who are members of your team, those who provide input but are not members of your team, those who are managers here to support your team, etc.

Ensure that all guests understand the length of your Kick-Off Meeting. These meetings are lengthy.

Ensure that all logistics are in place. The venue, meals, flip charts, etc.

5

HOLD THE KICK-OFF MEETING

Key points:

- Remain flexible and adapt to change.
- Focus on the outcomes, not the agenda.
- Get frequent feedback from the participants.
- Work with your Core Team during the meeting to ensure that everything works right.
- Enjoy it! Make certain that everyone else enjoys it!
- Recover all documentation at the end, including flip chart pages and notes.
- Schedule follow-up activities prior to the end of the meeting.
- Document all action items, and ensure that every one knows what they need to do after the meeting.

6

FOLLOW-UP MEETING AFTER THE KICK-OFF MEETING

The Core Team should meet shortly after the meeting, preferably the next day.

Key points:

- Recover all the documentation and other output from the meeting.
- Decide what to do next and who will do what.
- Disseminate all information quickly.
- Ensure that all action items are assigned and a delivery date is set.
- Clearly communicate the results of the meeting to all those who attended the Kick-Off Meeting. Also to the appropriate managers and executives.

LEARN AND IMPROVE (KAIZEN)

Debrief with your Core Team and learn from the experience. Solicit feedback from the participants.

Report to your management chain as appropriate.

Phase 2: Plan the Project

Example: Kick-Off Meeting Planning Tool

Name of the proposed Kick-Off Meeting: ______________________________

1. What is the purpose of the Kick-Off Meeting?
   What is the intent of the meeting? Said another way, why are you holding this meeting?

2. Purpose of this set of visits:
   For example:
   - ❖ To create full awareness of the scope of the program among the 25 team members.
   - ❖ To ensure that everyone knows their roles and responsibilities.

3. What are the tangible results that you must have from this meeting — i.e., the outcomes?
   For example:
   - ❖ A fully developed timeline.
   - ❖ Documented roles and responsibilities.
   - ❖ The formation of a Core Team.

4. What are the methods you will use to achieve the objectives of the Kick-Off Meeting?
   For example:
   - ❖ Create a very detailed agenda.
   - ❖ Ensure full participation by all team members.
   - ❖ Divide into groups to identify barriers and obstacles.
   - ❖ Create a Core Team to steer the program to completion.

5. What are the roles for the meeting?
   For example:
   - ❖ Who is the facilitator?
   - ❖ Who will coordinate logistics?
   - ❖ Who will take notes?
   - ❖ Who will create the handouts?
   - ❖ Who will make presentations?

Important: This form should be used at the earliest possible time prior to the meeting itself. Planning ahead is critical for a successful Kick-Off Meeting.

Important: Items 1-3 are the most critical elements of this form.

Ensure that you develop these items in collaboration with your team and managers.

Write objectives down and ask others to provide input. Ask an independent third party to provide input.

Finally, strive to answer all three questions. Time spent in this area will have a significant benefit to your success.

For comments, questions or additional copies please e-mail the author: flexpm@rapidinnovation.com.

6. What will happen the day after the meeting?

   For example:
   - All notes will be transcribed.
   - A meeting of the Core Team will be scheduled.
   - All action items will be sent to the attendees.
   - Who needs to know the results of the meeting?

7. What will happen a week or two after the meeting?

   For example:
   - The Core Team will meet.
   - Various sub-teams will have meetings.
   - A presentation will be made to the executives.

8. How will we evaluate the success of the meeting?

   For example:
   - Debrief the attendees.
   - Create an evaluation form.
   - By the number of deliverables we complete.

9. Who must attend the meeting?

   For example:
   - Engineering.
   - Marketing.
   - Manufacturing.
   - Technology (R&D).
   - Quality.
   - International.
   - Suppliers.
   - Channel management; sales management.
   - Documentation: Advertising and PR.
   - Managers.
   - Company executives.
   - Consultants.
   - Motivational speaker.
   - Training department.
   - Customer support; service and repair.
   - Calibration and metrology.

Phase 2: Plan the Project

Typical Progression for the Development of the Project Scope

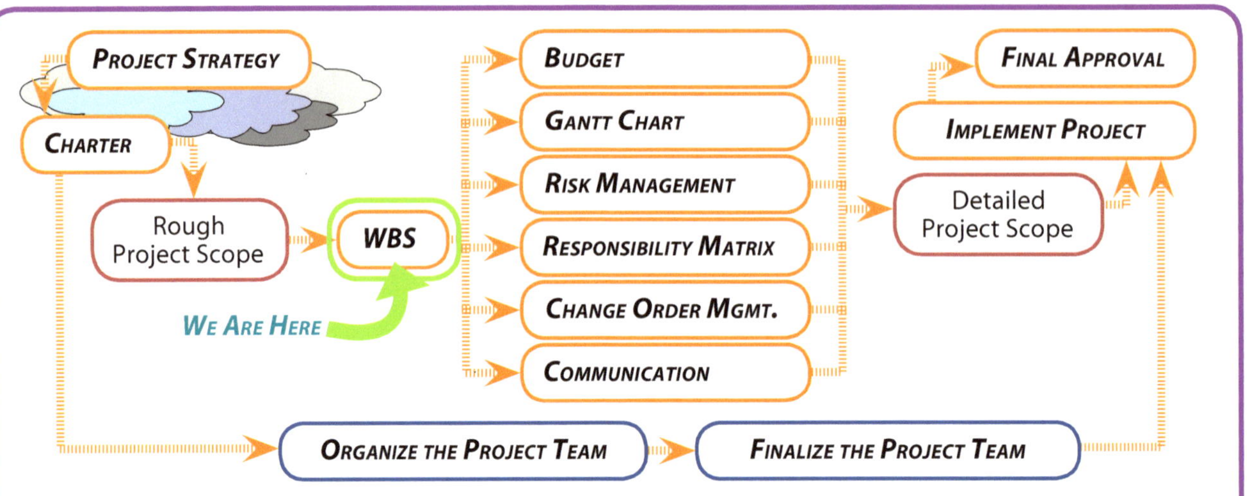

Work Breakdown Structure (WBS)

The Work Breakdown Structure (WBS)

A Work Breakdown Structure (WBS) is all about results. It is a way to display the project deliverables in a hierarchical fashion. The WBS is also called the Project Family Tree because the hierarchy starts with the overall deliverable and works its way down to the smallest deliverables. A WBS tells everyone what you will deliver — not how you will do it. These project tangibles tend to be physical in nature; for example, a completed building or a finished piece of furniture.

> Essential
> "A WBS tells everyone what you will deliver — not how you will do it."

In product development, the WBS might be called a Product Proposal or "roduct Plan. In software development, the levels (shown in the example on the next page) may show up as work blocks or work packages for the developers. On the other hand, as a project manager you must ensure that all the deliverables are identified and listed in the proper documentation.

Definitions

Bottom-Up and Top-Down

There are two distinct ways to develop a WBS. Top-down is generated by the project manager in consultation with the team. In some cases, companies have templates and documented processes.

The **TOP-DOWN** method is for more mature applications — for example, in construction where every project has common elements such as lay the foundation. Top-down also applies when there is much experience available and the what is relatively well-known.

BOTTOM-UP is more applicable where not much is known about the project — for example, when a team is doing a project for the first time or there are many unknowns such as technology, customer needs, etc.

Tree Structure

A tree structure is a way of representing a hierarchy in a graphical form; so named because the graph resembles a tree. The WBS is a tree structure that defines the work and deliverables in a related structure. The relationship is among each of the work elements and also between the work elements and the end product. Therefore, a WBS is a complete, hierarchical tree structure of tasks and deliverables that need to be performed in order for a project to be completed.

Phase 2: Plan the Project

The tree structure of the WBS helps you document all the deliverables of your development project. It starts with Level 0, the final deliverable — in your case, the new product. The next level represents the major elements, generally fewer than five, that, when aggregated, make up the new product (i.e., Level 0). Add levels as needed. There is no limit to the number of levels, but we must caution you not to breakdown into too many levels. Each level is made up of work packages, that is, complete deliverables.

Tasks

Tasks are part of a set of actions that accomplish a job. It is a set of value-added steps, the aggregate of which completes a job or deliverable.

What about CWBS, BOM and OBS?

There are a number of other acronym-based tools — for example, **CWBS** (Contractual Work Breakdown Structure) is used to define the level of reporting that the project contractor will provide. The **BOM** (Bill of Materials) is a representation of the physical assemblies, subassemblies, components and parts needed to manufacture a product. An **OBS** (Organizational Breakdown Structure) indicates which organizations are responsible for which work element from a WBS.

Clarification

There is a distinction between defining the project to develop a new product and defining the new product. The project includes all the activities, functions and tasks that are needed to get the new product to market and in the hand of your customer. Defining the product is all about the functionality of the new offering; i.e., value creation. The product definition is part of the project, but there is much more to the project than the definition.

Example: Work Breakdown Structure (WBS)

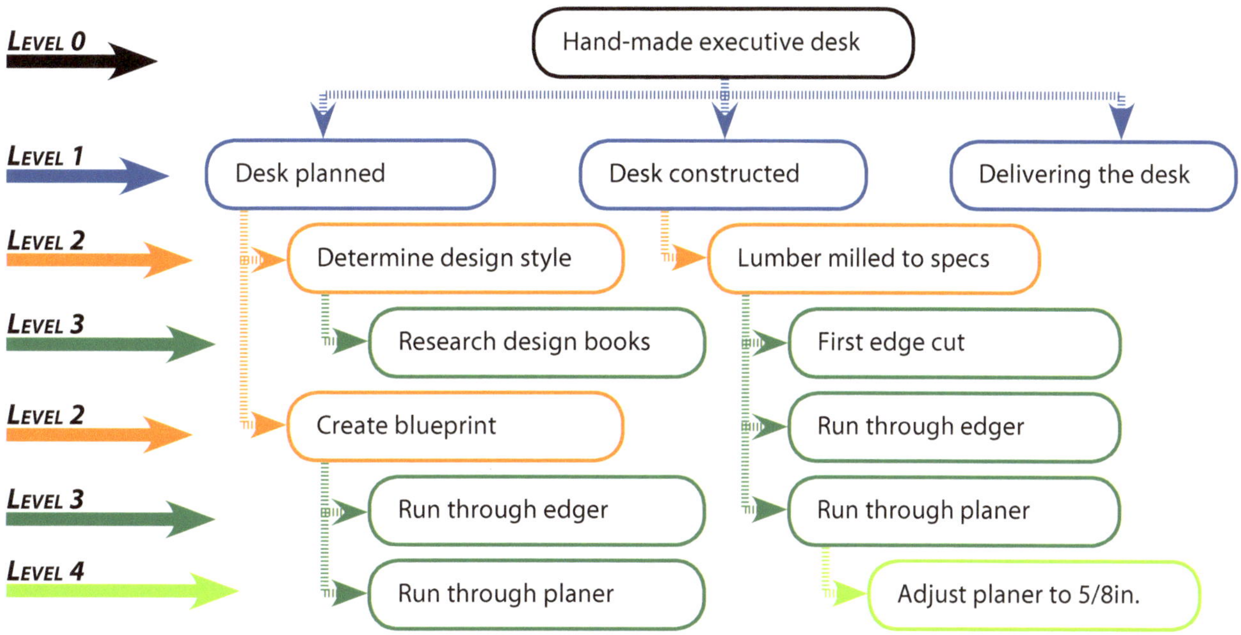

Here's a simple WBS to construct a desk. Notice the various levels. As project manager, you are responsible for leading Level 2 thru Level 0 (which, by the way, is the final product). The lower levels aggregate to the achievement of the Level 0 objective. Engineering leads, manufacturing managers, marketing people, et al manage the lower levels — they're in a better position to breakdown Level 3 and below.

Phase 2: Plan the Project

Detailed Process: Work Breakdown Structure (WBS)

1

Collect Needed Information

Identify the sources of information that will help you in developing the WBS.

For example:

- The project charter
- The project scope
- The voice of the customer
- Available resources — e.g., funding, people, equipment, etc.
- Assumptions, barriers, obstacles and other constraints Risks and other threats

Ensure that your WBS is focused on the deliverables.

2

Select the Type of WBS You Will Use

Three main types:

- Life cycle (chronological sequence).
- System (key parts or elements).
- Geography (key sites or locations).

Important

Too many levels may cause problems. The more levels, the more difficult to track and more overall complexity.

Use only as many levels as you judge needed to properly manage the project and retain control, and no more.

3

Determine the Level of Detail Needed for Your WBS

Key questions for this step:

- How complex is this project?
- How new is this project?
- How strategic is this project?
- How large is the budget?

4

Develop the WBS with Your Team

1. Identify major deliverables of the project based on the type of WBS you selected (life cycle, system or geography).
2. Next, start placing "Level 0" at the top. Pay careful attention to the phrasing or wordsmithing of Level 0 — it must be clear to all team members.
3. Divide major deliverables into smaller deliverables, and continue until you reach the lowest level.
4. Place "Level 1" in your diagram.
5. Continue to break down the work into ever-smaller work packages.
6. Place "Level 2" and all subsequent levels in your diagram.
7. Be careful not to define too many levels — rather than helping you better manage your project, it may add complexity that you don't need.

Important

It is best to use yellow stickies to develop your WBS, as you can easily move and reorganize them.

Use different color stickies for the various levels of your WBS — for example, blue for Level 0, red for Level 1, and so on.

Place the Level 0 sticky first. That way you can always glance at the ultimate goal.

5

Double-Check the Structure

Because there is no single correct WBS you need to double-check.

- Have you included all the deliverables?
- Do you have enough levels to properly manage the project?
- Can the information in your WBS be understood by all?
- Are the levels sequenced properly — i.e., lower levels build up to the higher levels?

Continued

For comments, questions or additional copies please e-mail the author: flexpm@rapidinnovation.com.

Phase 2: Plan the Project

CONT'D

6 DOCUMENT AND ARCHIVE

Document your WBS, as you will use it frequently.

Archive a copy just in case you lose the original.

Remember that many others in your organization will be using your WBS, so ensure clarity and ease of use.

7 USE THE WBS TO DRIVE YOUR PROJECT

For example:

- You can use levels 0,1 and 2 to make presentations to management.
- You can place a WBS on the wall to allow team members to study it and use it.
- Use it to organize your work on a daily basis.
- Use it for status meetings.
- Use it to develop your responsibility matrix, budget, risk management, etc.

LEARN AND IMPROVE (KAIZEN)

As you use the WBS, you will find ways to improve its accuracy, clarity and usefulness.

Your WBS will change over time; consequently, you will need to keep it up to date and relevant.

Examples: Work Breakdown Structure Numbering Systems

Example 1

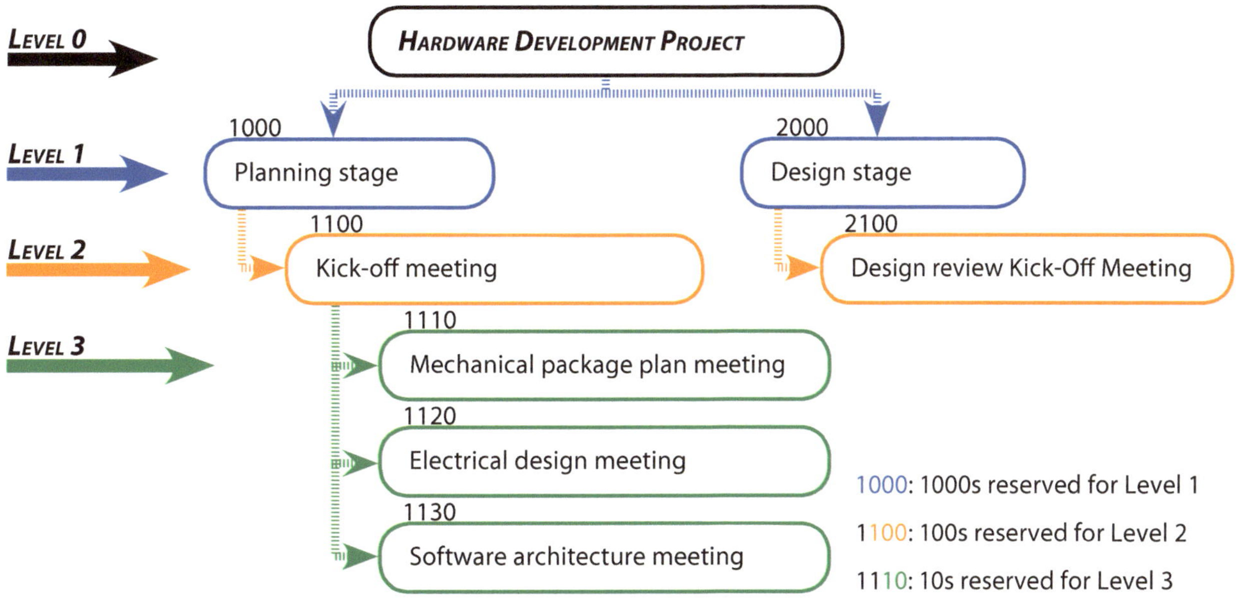

Phase 2: Plan the Project

Example 2

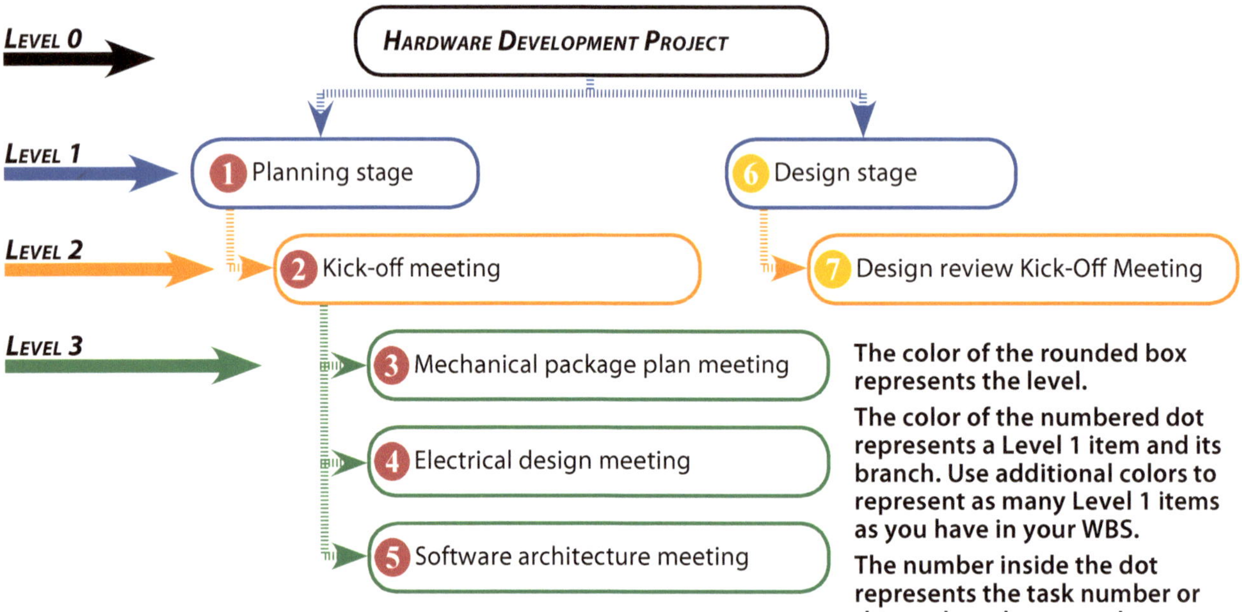

The color of the rounded box represents the level.

The color of the numbered dot represents a Level 1 item and its branch. Use additional colors to represent as many Level 1 items as you have in your WBS.

The number inside the dot represents the task number or the work package number.

Example 3

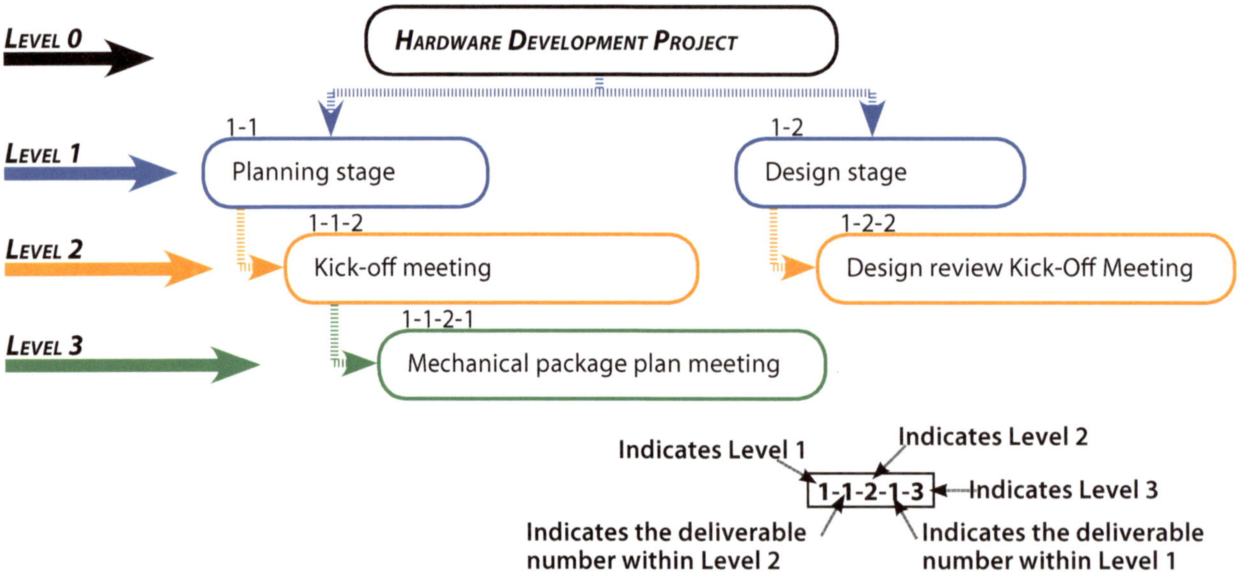

Phase 2: Plan the Project

Hierarchical Work Breakdown Structure

Many projects are large and complex and require the delegation of the WBS to various levels of project managers or functional managers.

The hierarchy approach you use depends on the nature of your project. For example, in product development, the delegation may occur chronologically — specifically, one project manager is assigned the planning, while a different one handles the development phase. Another choice is using the levels of your WBS. For example, levels 0 and 1 are led by one project manager. Lower levels are assigned to other project managers, functional managers and even team leads.

Important! Handoff levels need not be done exactly across your WBS. An asymmetric handoff strategy may be prudent. It depends on complexity, risk, organizational culture and other variables. Don't hesitate to mix levels if it is the best way to deliver results.

Critical Success Factors for a Hierarchical WBS

1. Ensure that there is a single project manager who oversees all other managers.
2. Clear communication is paramount — the complexity demands it. You must develop a comprehensive and robust communication plan. This might include regular status meetings, a Web site, newsletters, face-to-face meetings, etc.
3. The handoff points — when one manager interfaces with the level of another manager — are critical. You must develop a clear protocol to ensure clear execution. For example, what are the criteria for a handoff? Whose calendar do you use? How will conflict be handled?
4. Roles and responsibilities are of paramount importance, so ensure that everyone knows the handoff points and the criteria that constitute a handoff. Note that for every handoff there are at least two critical individuals: the sender and the receiver — the person who is delivering and the one who accepts it.

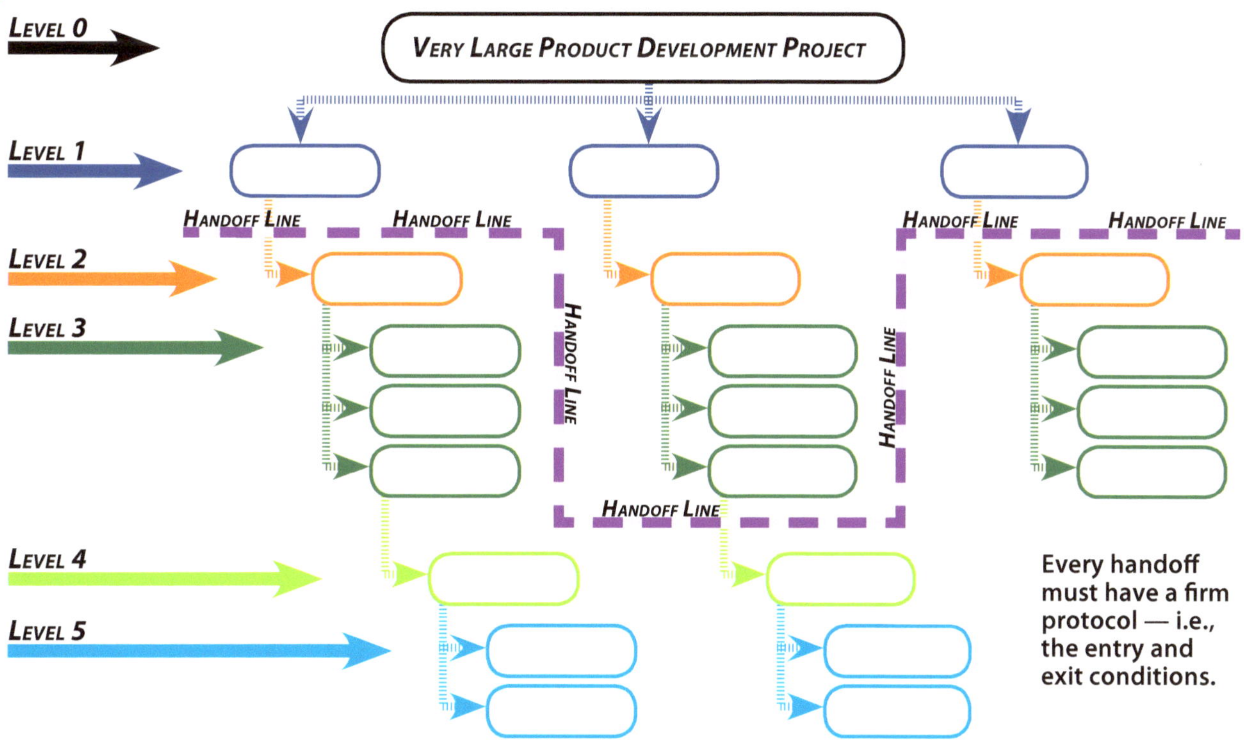

A single project manger should not manage all the levels of a WBS. It may not be physically possible. Also it suboptimizes the model. It is better to allow mangers and leads to manage the lower levels — they understand much better what needs to happen and by when.

Phase 2: Plan the Project

Typical Progression for the Development of the Project Scope

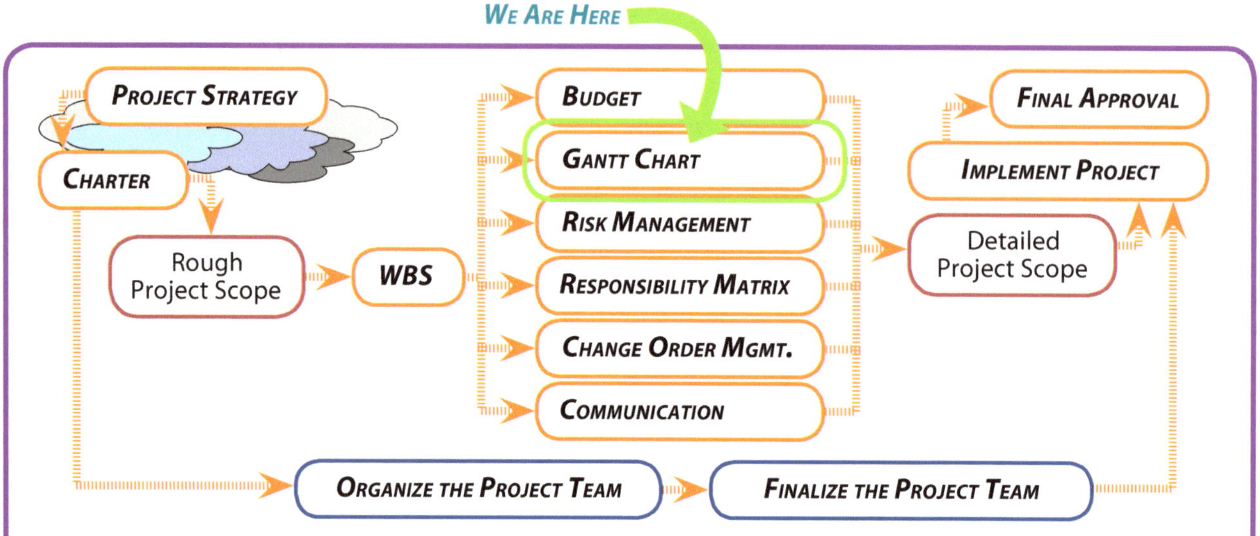

Gantt Chart

Scheduling with Gantt Charts

A Gantt chart is one way to represent the timeline or schedule of your project. The Gantt chart was developed in the early 1900s. It shows the project and each activity start and end against a horizontal timeline. It is widely used because of its simplicity. It is also an excellent tool to communicate your project's timeline to others on the team as well as managers and supervisors.

Scheduling and Product Development

Accurate schedules are probably one of the most difficult challenges in managing a development project. You have to deal with the complexity of the development itself. Plus the emotional component of those who create the timeline, under pressure from managers and executives to reduce time-to-market.

You must work closely with the engineering managers and engineering leads to obtain clear and realistic information. This calls for developing a high level of trust between you and engineering. You must also work closely with the marketing manager or product manager to determine the expectations for delivery of the new product. These two organizations tend to be at odds, engineering wants as much time as possible to design a good product; marketing wants it as soon as possible to compete in the market place. As the project manager, you must reconcile both forces and help determine a realistic schedule.

For your scheduling to work properly, there must be a high level of trust and empowerment in the organization. Your managers and executives must be understanding of the challenge (use the Complexity Diamond to communicate the challenge), but also set aggressive dates to compete and achieve business objectives. Engineering, Marketing and Manufacturing must understand the need to get the product to market and collaborate to achieve aggressive targets, without jeopardizing the quality of the new product.

Ingredients for Accurate Scheduling

- Schedules are not to be used as punishment. Schedules are honest attempts to predict the future, and require the full collaboration of all the functions to contribute to achieving the milestones. It is not one department against the other.
- There must be a single Master Schedule created and maintain by the project manager. Proliferation of schedules leads to confusion.

- Changes to the Master Schedule are not trivial activities. Changes must be made deliberately and with the full understanding of all the key stakeholders. Again, it's not one department against another, but the entire organizations working towards a common end.
- The Master Schedule must be readily available to everyone 24/7. It must not be a secret nor the domain of a few individuals. Naturally, common sense must be used to protect of confidential information.
- Openness, sincerity and trust are essential ingredients to creating and maintaining an accurate schedule.

Schedule Accuracy and Need to Know

Everyone in your team wants to know when the new product will be announced, even before you start the development project. This is normal, as the market place is competitive and turbulent. The marketing and sales organizations are eager to compete using the new product, your managers and executives want to start earning profits from the new product.

> Important
>
> Scheduling innovation is rarely possible. This is the one item that makes product development different.

It's a known principle of project management that as the time passes the accuracy of your timeline will improve. Yet, everyone wants to know very early in the development project, and with high accuracy. You run a risk that any date you set at the start of your development project is not likely to be accurate; said another way, there are too many unknowns at the start of the project to provide an accurate forecast of the completion date (NPI, New Product Introduction). For example, the technology you are using may prove much more difficult to implement thus delaying the new product. Your supply chain may be interrupted due to civil strive or natural disaster.

What to Do

Your team is entitled to know the Master Schedule so they can plan their own actions and deliverables. We suggest a three step approach:

1. **USE THE COMPLEXITY DIAMOND** to communicate to every one how big a challenge you have and where the hurdles are. This will help develop understanding of the complexity of the development project and thus manage expectations when it comes to schedule accuracy.
2. **USE PROACTIVE RISK MANAGEMENT.** Risk Management is all about managing uncertainty, things that can interfere with your schedule. When you neglect risk management, you allow uncertainty to manage your project. Risk management will provide the entire team with a better understanding of the challenge at hand and the ability to help manage the schedule.
3. **DEVELOP A DEGREE OF CONFIDENCE SCALE.** For example, at the start of your project, you may only have a 30 percent certainty of achieving the dates in your schedule. Once the project is progressing, your Degree of Confidence will increase. For example, a few weeks into your project your confidence level may be 60 percent, and so forth. Whenever you present the schedule be certain to include the Degree of Confidence and ensure that everyone understand how much they can rely on the information. It's not very difficult to develop a scale that you can use to communicate your schedule. Work closely with engineering and marketing — they will help you and at the same time develop a clear understanding of the meaning of the scale. When developing the scale you might want to use the Complexity Diamond — from it you can extract three to four criteria to use in developing your new scale.

There are more complex tools available to estimate the accuracy of schedules; for example Monte Carlo Analysis. While useful in some cases, their complexity may be counter-productive in your situation.

Phase 2: Plan the Project

Example: Gantt Chart and WBS

Timeline for a New House

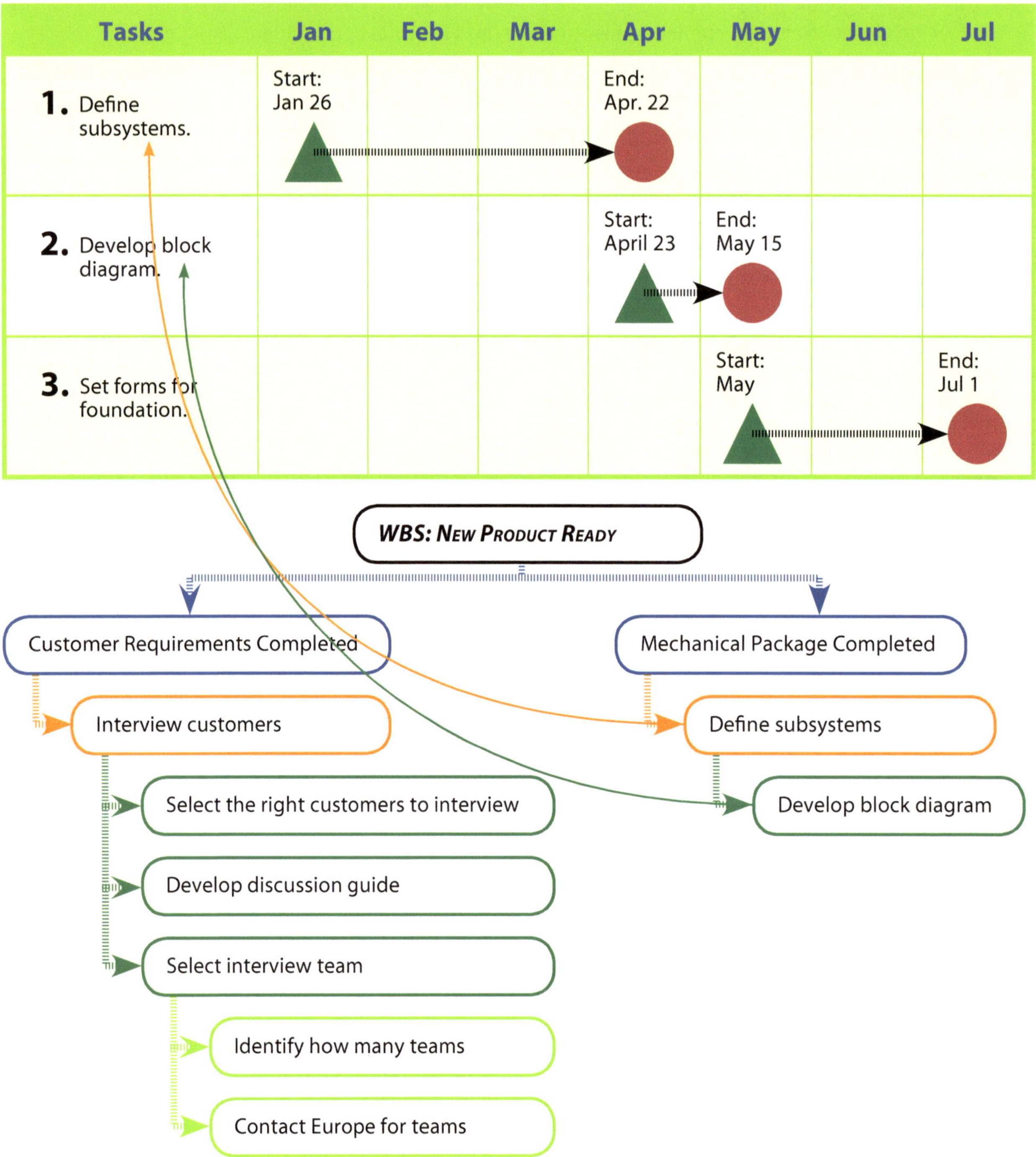

In the above example, the Gantt chart shows you see the relationship between the WBS and the schedule. You must have the WBS before you can develop your schedule. That being said, it is acceptable to make the development interactive as you may not have all the information ready at first.

Phase 2: Plan the Project

Detailed Process: Scheduling (Gantt Charts)

1

Collect Needed Information

Identify the sources of information that will help you in developing the Gantt chart.

For example:

- Project charter.
- Project WBS.
- Responsibilities.
- Contracts and agreements.
- Other relevant information.

You must use your WBS as the source of all the deliverables that need to be mapped in your Gantt chart.

2

Determine the Level of Detail

You need to determine how much or how little detail you need. For example, for a large project you may want much detail. There is no rule on the level of detail.

Use your WBS for guidance.

The length in months will also provide guidance — for example, how much detail do you need for each week of duration?

The number of people involved in your project can also influence the level of detail. The more people, the more detail.

Important! It is almost impossible to plot the information in the Gantt Chart prior to the sequencing, and duration (steps 3 and 4).

3

List and Sequence the Activities

Using your WBS, your experience and help from others, determine which activity must be done first, second and so forth.

Sequencing is very important, as it simplifies communication. All activities must be sequential in time — i.e., those that need to be done first will be shown earliest in the Gantt chart.

At first your calculations may not be exact; this if fine. As the project progresses you will add more and more accuracy to the duration and sequence.

4

Estimate the Duration of Each Activity

Duration is the calendar time an activity will take.

How long will each activity take? One day? One week? These are the key questions for this step.

Your experience, help from others and the history of projects in your organization will help.

Determining how long each activity takes in number of hours will help you estimate duration. For example, "10 hours to excavate."

Next, translate hours into days and factor in other variables such as scheduling, delays, etc, and you will have the true duration. For example, "10 hours to excavate," may take a full calendar week because you need to schedule the work and account for expected and unexpected delays.

5

Resolve Conflicts as Needed

Check list:

- Check for interdependencies and conflicts.
- Check for the duration of each element. Did you allow sufficient (realistic) time?
- Did you miss any activity?
- Collaborate with your team. Note: this is a critical step to ensure better accuracy and reality.

Continued

Phase 2: Plan the Project

CONT'D

6

PLOT THE INFORMATION IN YOUR GANTT CHART

As you develop the sequence, and determine the duration you start plotting the information in your Gantt chart.

Plotting the information on your Gantt chart will take several iterations. Be patient, and constantly reconcile your chart.

7

REFINE YOUR CHART

You will obtain new information and more details that will add accuracy to your chart. Continue to improve the chart over time.

Keep searching in your WBS, work with others on your team and use your own experience to add the amount of detail and accuracy needed to get the job done.

8

COMMUNICATE

The real value of your Gantt chart is to add to the clarity of communication. Keep your team engaged in your chart, and ensure that all stakeholders are fully informed.

It is of little value if you are the only one who is aware of the schedule — use your Gantt chart proactively to communicate and motivate.

LEARN AND IMPROVE (KAIZEN)

As you use the Gantt chart, you will find ways to improve its accuracy, clarity and usefulness.

Your Gantt chart will evolve with the project. Continue to communicate with all stakeholders as the Gantt chart changes. Also, rely on the input of all stakeholders to keep the schedule relevant.

Example: Gantt Chart and WBS

Timeline for a New House

	Level From Your WBS			Months									
#	L1	L2	L3	Jan	Feb	Mar	Apr	May	Jun	Jul	Aug	Sep	Oct
1	Build a new house.			Start 1/26									End 10/15
2		Clear the property.		Start 1/27	End 2/25								
3			Survey the property.	Start 1/27 End 1/29									
4			Bulldozer clears.	Start 1/30	End 2/10								
5			Haul all the debris.		Start 2/1 End 2/11								
6		Build the foundation.			Start 2/10	End 3/10							
7			Survey the foundation.										
8			Lay the forms.										
9			Pour concrete.										

Phase 2: Plan the Project

Hierarchical Scheduling (Gantt Charts)

Introduction

As a project manager, you are responsible for managing to results — that is, you must be constantly focused on ensuring that your project delivers the anticipated results. As such, you are the keeper of the Master Schedule — the schedule that drives all other schedules in the project.

The larger your project, the more levels in your WBS and, consequently, the greater the challenge of scheduling. We recommend that you focus on managing at the higher level of the schedule and allow others to manage the smaller portions. This strategy will help you avoid being overwhelmed by too much detail, and it also keeps you from micro-managing.

A hierarchical schedule is a multilevel schedule with varying stages of detail on each level. Every activity in a higher level schedule is broken down into several activities (and sometimes entire schedules). Typically, the schedules from various levels are connected at major milestones or events.

The Master Schedule drives all other schedules. It is a milestone schedule and it is the responsibility of the project managers. All other schedules are the responsibility of other managers and even individuals.

Phase 2: Plan the Project

Example: Hierarchical Scheduling

The Master Schedule Drives Every Other Schedule

Tasks	Jan	Feb	Mar	Apr	May	Jun	Jul
1. Plan approved by the county engineer.	Start: Jan. 26			End: April 22			
2. Clear the property.				Start: April 23	End: May 15		
3. Set forms for foundation.					Start: May		End: Jul 1

Tasks	Jan	Feb	Mar	Apr
1. Initial paperwork.	Start: Jan.27	End: Feb. 27		
2. Meet with county officials.			Start: March 2	End: April12

Tasks	Mar	Apr
1. Prepare for initial meeting.	Start: March 2	End: March 7
2. Conduct first meeting.	Start: March 7	

Phase 2: Plan the Project

The Responsibility Matrix

Introduction

In all projects there are complimentary roles and responsibilities. There are the people who approve the work, the people who need to know about and review the work and the people who actually do the work. In small projects this tends to be very clear — for example, the manager reviews and approves. As projects get larger, the picture gets more complicated, and the need to use a tool becomes imperative.

The responsibility matrix is one way to define roles, responsibilities and authority. It's a way to manage expectations by overtly defining each person's role. Additionally, it communicates to others that they play no role in the project.

Responsibility Matrix and Product Development

Why a Responsibility Matrix

- It manages expectations by clear and early definition of roles and responsibilities.
- It avoids needless work by reducing the number of people who can obstruct forward progress.
- It is a good way to gain agreement of roles and responsibility by forcing the conversation — i.e., people engage in the development of the responsibility matrix; thus clarifying their roles.
- It addresses the complexities of a team that is not collocated.

Not many product development organizations develop a responsibility matrix. This is due to the familiarity of the entire development team, which has up to now been collocated. But this may need to change because of the emergence of multi-site development, co-development and outsourcing.

When the development team is not collocated, the need to document roles, responsibilities and authority becomes imperative. People in other sites may not be familiar with the structure of your organization, let alone people who work for other companies such as suppliers, partners and contractors. You should be able to develop and post a responsibility matrix on-line, making it easy for everyone to view.

Its value, however, is only realized when people actually use it. Seek the support of your managers to implement a responsibility matrix for your development program.

Definition of a Stakeholder

Technically, a stakeholder is somebody with direct interest in your project or program. For practical purposes this definition is not sufficient. Thus, we prefer to define a stakeholder as somebody who has a vested interest in the success of your project. For example, your manager, who will not use your service but does have a stake in ensuring success.

Stakeholders, then, are those persons or organizations that have something to win or lose depending on the outcome of your project.

Generally, a stakeholder is not a customer, but it can be someone who could seriously influence your project.

Template

Example: Responsibility Matrix

Project Name: "EL DORADO" Date: November 2009

Project Leader: Ralph Gooch

It's always helpful to add the name and title of the individual. It helps avoid confusion.

	Project Sponsor	Functional Manager	Project Manager	Project Team	Steering Committee
Project Definition.	(R) Fred Woo, Gen. Mg		(R)	(D)	(I)
Business Requirements.		(A) Mary Gomez, Mktn.		(D)	
Budgeting.		(A) Al Wolf, Prod Mgr.	(D)		
Status Reports.		(I)	(D)		
Risk Management.			(L)	(D)	

The codes you select need to be relevant to your project. Select only a few codes to simplify communication

Code	Responsibility
A	Approve
R	Review

Code	Responsibility
C	Create
D	Deliver
I	Inform

Code	Responsibility
IN	Provide Input
N	Notified
L	Leads a Team or Sub-Team

Here is a typical responsibility matrix. Use it to develop one that better fits your business objectives and the culture of your team. There are many variations of the above matrix, but they all mean to add clarity to the roles, responsibilities and authorities of the stakeholders in your project.

Notice the matrix at the bottom; it describes the various levels of responsibilities. This is a protocol that must be developed in full collaboration with your team and the managers of your organization. Its value is dependent on full understanding and support of the concept. You can modify the meaning or add new levels to better suit your needs — do not hesitate to customize. The first column simply lists the various activities or documents associated with your development project. You must develop this list to reflect the reality of your project.

Do not hesitate to place the matrix on a web site and distribute it to every member of your team. The concept only works if everyone is aware of the matrix and willing to support it.

Phase 2: Plan the Project

Examples: RACI

RACI is another form of the responsibility matrix. It relies on cross-functional responsibility rather than responsibility for each activity. While the standard responsibility matrix lists the activities and then assigns responsibilities, RACI identifies the function first, then the responsibilities assigned to that function.

Example 1: Based on Function

All activities belonging to a single department or function are listed.

Engineering					
No.	Task	R Review	A Approve	C Contribute	I Inform
1.A	Define specifications.	Fred Smith, Design Engineer	Mary Thompson, Engineering Manager	Bob Menlo, Engineer	Bob Chancellor, Manufacturing Engineer
1.B	Document circuit design.	Robert Ratzo, CAD/CAM			
1.C	Develop theory of operations.	Charles Metro, Documentation	Mary Thompson, Engineering Manager		

Example 2: Based on Task

All activities belonging to a single project or task are listed.

Pour the Foundation					
No.	Task	R	A	C	I
1.A	Create the plan.	Fred Smith, Design Engineer	Mary Thompson, Engineering Manager	Bob Menlo, Engineer	Bob Chancellor, Manufacturing Engineer
1.B	Purchase materials.	Robert Ratzo, Purchasing Department			
1.C	Place the forms.	Charles Metro, Carpenter		Mary Thompson, Laborer	

For comments, questions or additional copies please e-mail the author: flexpm@rapidinnovation.com.

Phase 2: Plan the Project

Typical Progression for the Development of the Project Scope

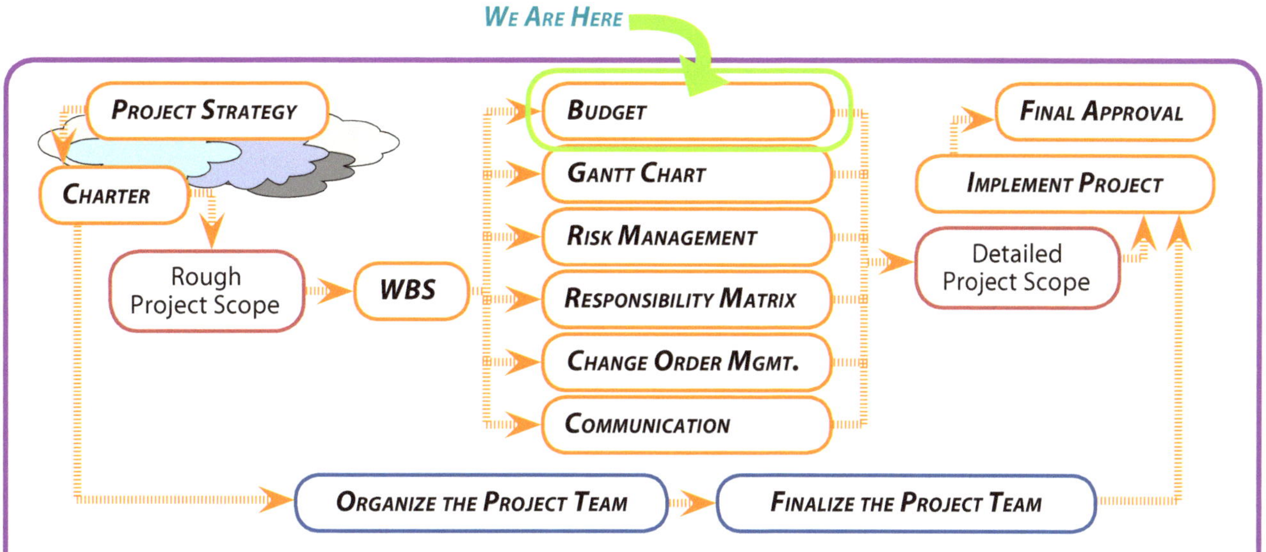

Budget

Budgeting and Cost Estimates

Estimating the cost of a project is a critical step to sound project management and a responsibility to the company or organization. It is one of the three key parameters for any project — the other two being scope and time.

Cost estimating can be a high-risk activity, yet it is very important. You need to dedicate time and resources to do it right. And you must monitor constantly to ensure proper financial management of your project.

Important! It's critical that you define your role as the Project Manager in the budgeting for the development program.

In many organizations each function such as Marketing, Engineering, Manufacturing develop and manage their own budgets for the project. Generally, you will be working with the budget developed for the engineering part of the development. Be certain to discuss the roles and responsibilities for budgeting with the functional managers.

Definition

A cost estimate is a process to calculate and document the resources that are needed to complete project activities. It is therefore the process of tabulating and predicting the cost of a project.

Why Estimate

A cost estimate increases the chances of approval and funding for your project. It also enables better measurement of performance and provides early detection of problems. Estimating also allows for comparison of alternatives and the cost of major changes.

Key Terminology

- **ESTIMATE (COST ESTIMATE):** Tabulating and predicting the cost of a project.
- **BUDGETING (COST BUDGETING):** The process of translating cost estimates into budgets.
- **CASH FLOW (FORECAST):** How and when the money will be spent over time.
- **PROJECT BUDGET:** The amount and time-phased distribution of money allocated to a project.
- **COST CONTROL:** The aggregate benefits of the estimate plus the budget plus the cash flow.

Phase 2: Plan the Project

Budgeting and Product Development

Developing a new product may cost millions of Dollars. Budgeting involves critical, strategic decisions. Your company may already have a process to approve new product development. If so, we suggest that you depend on this process. The decision to proceed with a new product development is made at the high levels of the company. As a project manager you may be asked to submit estimates of costs that will be added to the overall investment in the new product.

You will need to collaborate with engineering, marketing, manufacturing and other functions in order to develop your section of the overall budget. Become familiar with the budgeting process in your organization — it varies by company and even by function.

The critical challenges in product development are time-to-market and customer requirements. Budgeting some times takes a backseat to the previous two. In a competitive environment timing is critical, so it takes precedent over budgeting. So we say, become familiar with the budgeting process in your company!

Step 1: Define Why You Need to do an Estimate

It's easy to get sidetracked by the urgency of developing an estimate and bypass the most important question: Why are we doing it? The first step in any estimate is to define its strategic value — i.e., the intent of the estimate. You should meet with the project sponsor, managers and other key stakeholders to identify the real purpose of the estimate. There must be clarity of intent if you are to succeed in your efforts.

Assess Capital Cost

- Validate capital spending — for example, purchase of expensive equipment.
- Define amount to borrow for financing.

Evaluate Risk and Productivity

- Identifying the likelihood of project change and, thus, changes in the cost.
- Preparation of a bid or a proposal.
- Develop a contract.

Establish a Baseline

- Estimate the resources — for example, the hours needed to complete the project.
- Compare with other bids or estimates.
- Build confidence that your company can assume the burden.

This diagram shows three distinct reasons for creating a budget for a new product, and there are others. It is critical that you understand the reasons why you were asked to produce a budget; otherwise, you may not meet the needs of your organization. Thus, there are two questions you must answer before you get started. What is the reason or the intent of the budget? What is the process your organization uses to create and present a budget?

Accuracy of Budgets

When you start a project the accuracy of budgets tends to be low, prone to errors based on assumptions. We strongly suggest that you attached a Level of Confidence Percentage to your budget estimates. For example, at the start of your project the accuracy might be ± 30 percent. As you develop better understanding and new data arrives, your accuracy will improve.

The risk to you is that the first number becomes the official budget; thus, causing you to have to deal with an unrealistic budget. Develop a clear agreement with managers that the accuracy will improve over time. Also, impress on the managers that the first estimate is just that. Finally, add the Level of Confidence Percentage to every budget you submit.

Example: Analogous Estimate

Cost Estimate

Company Name: ________________ Project Number: ________________ Estimate Number: ________

________________________________ Project Manager: ______________________________ Date: ________

Line Item	Type of Labor						Total Cost of Labor	Overhead	Billable Cost
	Engineering Rate: $100		Technician Rate: $50		Helper Rate: $30				
	Hours	@ Rate	Hours	@ Rate	Hours	@ Rate			
Develop schematic.	12	$1200	22	$1100	15	$450	$2750	10%	$3025
Visits to the site.									
Develop bill of materials.									
Wiring and assembly.									

Example: Develop the Cash Flow or Time-Phased Budget

The cash flow is a forecast of how and when the money will be spent. This is a critical step as it allows you to measure and monitor the cost performance of the project. It is essential in managing the project. In some organizations this is called the burn rate, which is rate of expenditures versus time. Companies need to know not only how much you will spend on your development program, but also at what rate; i.e., when the expenditures will take place over the life of your project.

Cash Flow

Company Name: ________________ Project Number: ________________ Estimate Number: ________

________________________________ Project Manager: ______________________________ Date: ________

Line Item	Jan	Feb	Mar	Apr	May	Jun	Jul	Aug	Sep	Oct	Nov	Dec	Ttl
Develop schematic.			$4400	$50	$100								$4550
Visits to the site.	$4000	$3000											$7000
Develop bill of materials.					$300	$700	$1200						$2200
Wiring and assembly.								$3000	$2500	$50			$5550

Phase 2: Plan the Project

The Business Case

Frequently, managers and executives ask you to make the business case prior to approving a new project or the budget for your project. This is a valid request that needs to be addressed with energy and due diligence.

Making the business case is communicating the investment needed for your project and the return on investment in a compelling fashion that includes clear financial information. It is also your conviction that the investment will improve the productivity of your company or organization.

Remember that managers and executives speak the language of dollars, as they have the responsibility to protect the interests of the company owners and stockholders. After all, the company exists to deliver a profit. Managers have the obligation to ensure that every investment furthers the profitability of the enterprise. Your obligation is to ensure that your project is communicated in the same language that management speaks — primarily, the language of dollars.

The business case simply means that you have done the proper research to ensure that the investment to be made will provide a vivid and defensible return. Defensible in this case means clear and based on a solid relationship to the profitability of the company or organization.

General Rules for Making the Business Case

- Keep it simple. Remember that your managers want to hear only three things:
    1. I need this much money for my project.
    2. When it's completed it will provide the following return on the investment.
    3. These are the risks involved in this project.
- Managers look for three specific types of returns (which all contribute to the bottom line):
    1. Improvement in speed.
    2. Improvement in customer satisfaction.
    3. Reduction of waste.
- Avoid terms like "I hope," "I feel" and "I think" because they do not convey due diligence or conviction. It is best to say, "I know," "I believe" or "I deem."
- Managers also look for investing the least for the most return; it is natural and part of their responsibilities. Thus, the amount you request must be carefully thought out. Ask yourself: Is this the least expensive way to complete my project?
- Remember: Opinions confuse; data convinces. Two things that annoy managers are vagueness and platitudes. It is better to say, "We will improve productivity by five percent," than to say, "We will have a massive improvement in productivity."
- Yes, everything can be translated into the language of dollars. Sometimes it may take special effort, but it can be done.
- Conviction is part of the challenge. Your managers don't have insight into your project, and thus rely on you to project conviction that you will deliver the expected results. Indeed, this takes courage and due diligence, but it is expected.

But How About Intangibles Like Morale Improvement or Goodwill?

Only if you can translate them into the language of dollars (sorry!). While intangibles are desirable, your managers have the fiduciary responsibility to the company to invest every dollar wisely and provide a return on the investment. Therefore, you must quantify intangibles in terms of dollars, improved time-to-market, profitability, reduced waste, etc.

Critical Success Factors (CSFs)

CSFs are those things that must happen in order to achieve the desired results. Consequently, CSFs can be described as those things within your project that really matter for its success and therefore must happen. The leverage of CSFs is equally powerful to risks in managing your development program. In fact, CSFs are the converse to Risks.

The benefit of CSFs is the ability to focus your team, managers and key stakeholders on the critical items needed to make your new product a success. You must involve your team in the creation, prioritization and review of CSFs. This creates a common set of references and helps the CSFs become integral to your program.

The process for developing the CSFs is the same as for Risk Management. First, you assess — that is, you identify and prioritize CSFs. Then you plan and implement. This means you plan preventive and corrective action (mitigate) and then implement preventive action. Finally you monitor and control, which also includes taking remedial or corrective action as needed.

The methodology to develop CSFs is the same as that of risk management — naturally, with the exception that CSFs must happen. You can use the same processes, prioritization, logs, protocols and other tools to identify and manage CSFs as you do risks.

We always recommend that your list of CSFs be short. You must prioritize them so you can manage them. Your list should be three to five items. Perhaps in a large development program, such as a brand new platform or products for a brand-new market, your list may double. But always keep it short and manageable.

To develop useful CSFs you must start with the business objectives of your development project — much as you do with risks.

There are Several Levels of CSFs

- **COMPANY:** The factors that are critical to the success of your company as they relate to your development project. In short, those things that must go right in order to support the business strategies of your company.
- **TEAM:** The factors that are critical to the success of your team; for example, the need to have a face-to-face meeting at the start of your project.
- **PROJECT:** The factors that are critical to the success of the project; for example, the fact that the public announcement of the product takes place on a very specific date.
- **PERSONAL:** Those things that you must accomplish in the development project to further your career and professional success.

There could be more categories of CSFs — you decide. On the other hand you need not use any pre-determined number of CSFs for any project. Use your judgment, work with your team and decide. We suggest that you keep the number of categories and the number of CSFs to a minimum in order to preserve the value of the concept.

Key Points

- There are several levels of CSFs: personal, team and company. Ensure that you consider all the CSFs at all levels.
- Critical Success Factors and Risk Management go hand-in-hand.
- It is crucial that everyone on the team must be involved throughout the entire process.
- Keeps your list of CSFs short — the three to five *critical* success factors.

Phase 2: Plan the Project

Examples

- The algorithm to render the new GUI must be tested and ready to use no later that the start of Q3.
- The sales organization must be trained on the new financing approach no later that Q4.
- The new display for the instrument must meet military standards nnn/nn/n.
- The testing of the GUI with potential users must be comprehensive and be concluded on time — no later than October 1, 20XX — to be ready for incorporation in the Alpha product. Fred S, is the owner and champion of this CSF.
- All major account visits must be successfully concluded by January 27, 20XX. May J. is the owner and champion of this CSF.

As you can see from the above examples, your CSFs can relate to any essential part of your new product. You and your team must collaborate to identify a short list, monitor the list and take corrective action if any item in the list is at risk.

The other side of Critical Success Factors (CSFs) is risk. We highly recommend that you address both categories in your projects. The methodologies are the same; that is, the processes are the same except that for CSFs you focus on those few things that *must* happen, while with risks you focus on things that *could* go wrong and *must not* happen.

Risk Management

Risk is the chance that something can go wrong with your project. Corollary: Ignoring risks does not make them go away! On the other hand, risk management is the disciplined process of identifying risks, evaluating them, preparing a response and implementing the response at the right time to avoid or reduce damage to your project. In short, risk management is answering the following key questions:

1. What can go wrong?
2. What is the probability of it happening?
3. What are the consequences if it does happen?
4. What can we do to avoid it happening?
5. What can we do if it does happen?

The Business Case for Risk Management

The business case for deploying a framework of risk management in product development is to achieve any of four specific goals:

1. Reduce time-to-market.
2. Improve reliability.
3. Reduce waste (improve efficiency).
4. Improve customer satisfaction.

Risk management is a way to improve efficiency in product development. Furthermore, risk management has a direct and clear impact on profits.

The diagram on the right defines several business benefits of risk management. In its simplest form, the business case for risk management can be made by the following assertion: Risk is always there; ignoring it is a flawed strategy that will not make it go away.

How many times have you heard: "That problem came out of nowhere."? We beg to differ; most "surprises" in product development are predictable, and can be mitigated before they become a big problem.

The most important reasons for doing risk management is to reduce time to market, improve customer satisfaction, improve reliability of your new product and reduce costs.

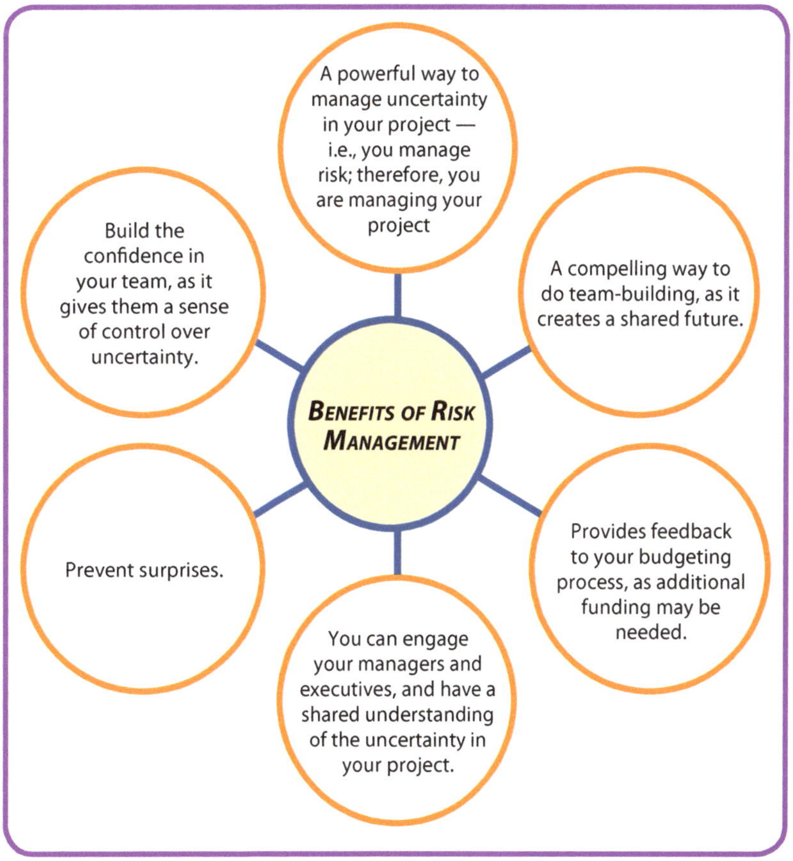

Phase 2: Plan the Project

The System for Closed-Loop Kaizen

Continuous Improvement at Its Best

This diagram shows the opportunities of combining Risk management with debriefs. Risk management is about predicting what could go wrong with your project, and debriefs are about analyzing what went right and wrong with your project. It stands to reason that the two should be connected. Think of them as the bookends for every project.

> **Important**
>
> **Assess the Effectiveness of Risk Management**
>
> Applying a model of risk management to your development model is a very good first step. Next, you must assess if the model you have implemented is improving your development ability. We highly recommend that you conduct debriefs (also called postmortems and retrospectives) on your development project, with the specific objective of evaluating the risk management model.

START HERE

Step 1: Plan

RISK MANAGEMENT: PREVENTION AND MITIGATION OF DEFECTS

Risk Management	FMEA
Risk management prevents defects from occurring and mitigates if defects do occur.	FMEAs are **ADVANCED** forms of risk management.

Step 2: Act

PROJECT EXECUTION

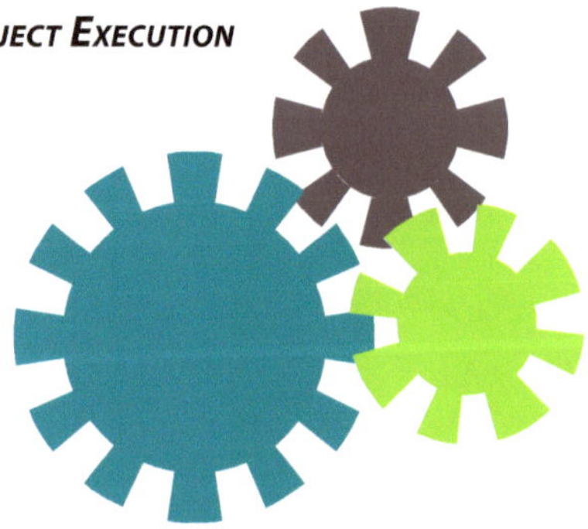

Step 3: Measure and Review

DEBRIEFS: CREATING THE CAPACITY FOR FUTURE IMPROVEMENT

- Debrief only selected projects
- Focus on clear action planning and execution

Step 4: Improve

BEST KNOWN PRACTICES

An auditable process to prioritize, archive and disseminate good practices, which will then drive measurable improvement in the organization. This process is generally provided by the Project Management Office.

Process Points

- Acquisition of information.
- Processing and prioritization.
- Archiving.
- Dissemination.
- Measuring and improving.
- Cleansing and archiving.

Types of Risk

A Few Examples

There are many types of risks. We encourage you to look at the big picture of your project and avoid tunnel vision. Risks include everything that might interfere with the delivery of the objectives of your project. Some risks will be unique to your industry, your company and even the country in which you reside. Use the following sample as a starting point, continually improving it with risks more relevant to your situation.

Types of Risk	Examples
1. Political (Internal and External)	• Personal agendas create obstacles to your project. • Two managers compete for a promotion; your project is caught in the middle. • Government or industry regulatory mandates interrupt your project. • New safety regulations cause a long delay for your project.
2. Cultural	• Your project involves a cultural change in the organization, causing obstacles. • A new acquisition means you must deal with two (corporate) cultures: that of your own company and of the new company. • Conflict arises because your team is scattered in across several countries. • The CEO of the company is risk-averse and considers your project too risky, while you believe it to be very manageable.
3. Economic	• A recession causes budget cuts and layoffs. • Added costs make your project unprofitable.
4. Technical	• The technology you are using in your project does not work. • The technology you are using creates a safety hazard.
5. Legal	• The owner of the IP you are using sues your company for infringement. • A restrictive law in another country will impact the IP you are planning to use.
6. Weather	• A sudden snowstorm brings your project to a halt. • An earthquake damaged the facilities of one of your suppliers.
7. Resources	• Your key engineer takes ill and is unable to continue working. • Two of your top scientists are lured away by a competitor. • A layoff could reduce your team.
8. Organizational	• A reorganization is announced in the middle of your project, causing you to lose several engineers to a new division. • Your division has no method to prioritize projects — every project is "Priority 1." • Your company is acquired and a new culture emerges, thus creating disruptions.
9. Procurement	• Technology you wish to purchase is unprotected and available to competitors. • The factory that you wish to use to manufacture your new product has gone on strike and there is no end in sight.
10. Market	• A new industry standard is enforced, which may cause a product redesign. • A new technology is being adopted by your competitors, which may cause redesign or a complete new design for your products.

Phase 2: Plan the Project

Process Map: Risk and the Product Life Cycle

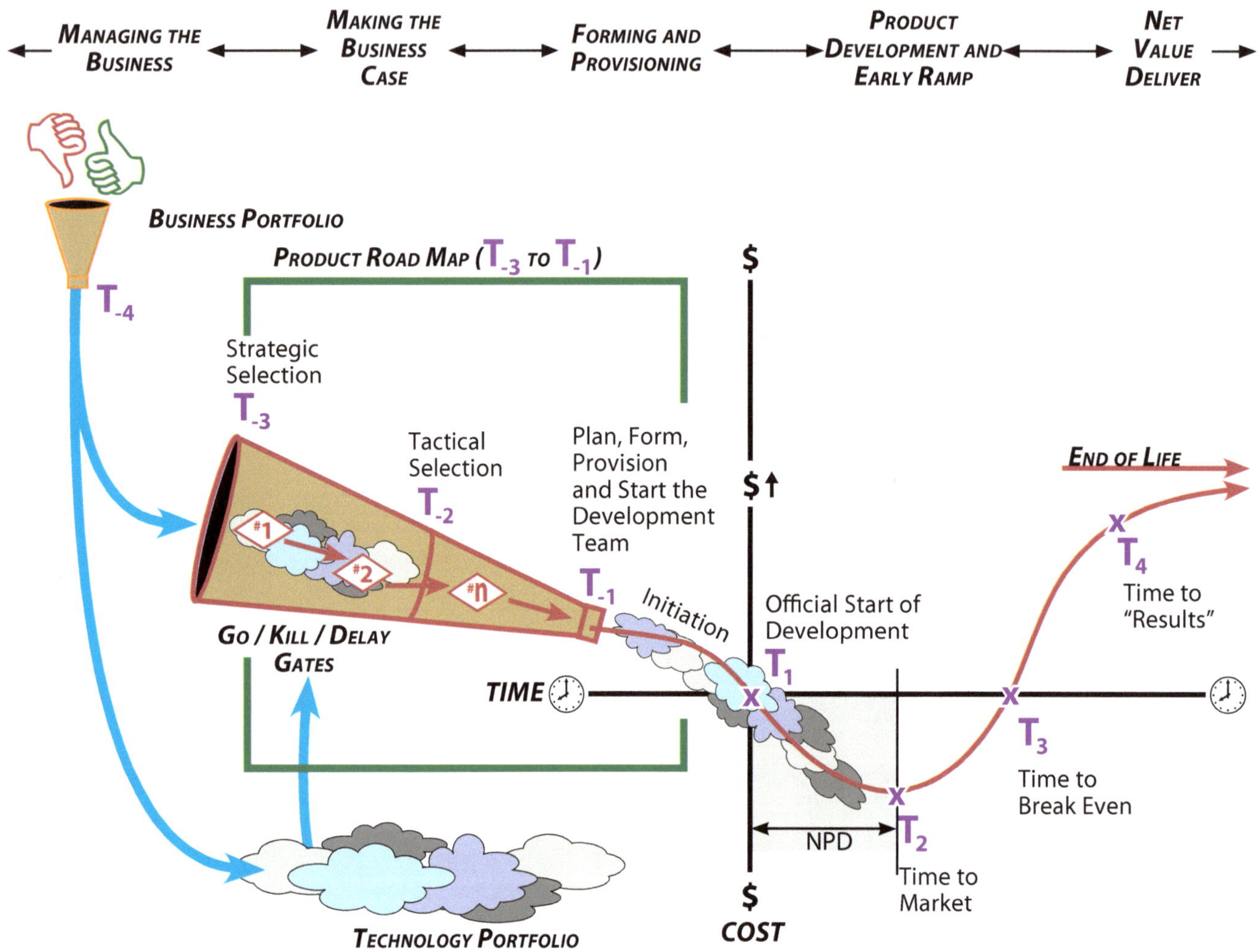

Managing the Business	Strategy Implementation	Tactical Selection of Project	Product Development	Product Introduction	Maturity
• Risk in the economy. • Macro risks. • Risk of not understanding the marketplace. • Risk of poor selection of technology. • Risk of poor portfolio management.	• Technology risks. • Market risks. • Risk of competition. • Risk of a disruptive new product.	• Risk of prioritizing the wrong project. • Resource risk.	• Risk of not understanding customer needs. • Resource risk. • Supplier risk. • Manufacturing risk. • Quality risk. • Materials risk. • Reliability risk.	• Risk of a poor introduction to the market. • Risk of the product not being ready to sell. • Risk of poor support. • Risk of a fatal bug.	• Supplier risk. • Manufacturing risk. • Competitive risk.

The above mental model is a depiction of the product development system. It shows the start at point T_{-4}, where the corporate decisions are made to manage growth and profitability. Note point T_1, which shows the start of a typical product development project. Also note that the New Product Development Process (NPD) starts at T_1 and ends at T_2. You must take into consideration the entire system to develop the right risk management plan at the right time.

Understanding Risk Management

Risk, Chance and Gambling

A critical step in the implementation of risk management is the clarification and adoption of a uniform language across the organization. People have different perceptions or understandings of the terminology used in risk management. It is important for your organization to address these differences and develop a unified set of definitions to avoid misunderstandings.

> **Important:** Every organization should create a one-page glossary where key risk management terms are clearly defined. Many people have preconceived interpretations of various terms. A company glossary will go a long way toward aligning everyone to a single meaning; thus, avoiding confusion.

- **AMBIGUITY:** Doubt about meaning and lack of clarity; something that can be understood in more than one way. For example, the behavior of key stakeholders in your project. Ambiguity is related to uncertainty.
- **CHANCE:** Related to uncertainty, it implies the willingness to make decisions in the absence of information. Good for gambling, not for project management. Chance is when you don't know the probability of an event.
- **GAMBLING:** Not the same as risk management because gamblers can choose their risks, while in business the risks are there already, waiting to be managed.
- **MITIGATION:** Anything that counters the negative effects of the risk or the sources of the risk; e.g., prevention, contingency planning and reserves.
- **PREDICTABILITY:** The degree of something that happens as expected; to tell in advance with accuracy.
- **RISK:** The probability of something going wrong with your project. The risk is known and one can mitigate it, avoid it or remedy it. Unlike chance, risk is knowing the probability and then managing with the information you have at hand.
- **UNCERTAINTY:** The state of being in doubt about something; the estimated percentage by which an observed value may differ from the true value.
- **VARIABILITY:** Something that is unpredictable, able to change and not constant; for example, the budget for your project varies due to the economy, currency fluctuation or budgeting process. Variability is also the degree to which something varies — the more variability, the more possible outcomes.

Glossary of General Risk Management Terms

Although this is a useful glossary of general risk management terms, this is not an FMEA glossary.

- **RISK PLANNING:** The decision-making process to define the approach (strategy) and plan the risk management activities for your project.
- **RISK IDENTIFICATION:** The process to determine and document which risks will likely affect your project.
- **RISK PRIORITIZATION:** The process to assign levels of importance to the various risks that are possible in your project, based on clearly understood and written criteria.
- **RISK MANAGEMENT PLANNING:** The process to develop the reduction of threats to your project by means of procedures, techniques, roles and responsibilities.
- **RISK MONITORING:** The process to detect early signs of risk through the life of your project.
- **RISK CONTROL:** The process to execute the risk reduction plans to mitigate or eliminate the damage of risk to your project.
- **RISK MITIGATION:** The steps you take to prevent, correct or deal with risk. Generally, made up of two parts: the steps you take to avoid the risk from ever taking place and the steps you take if it does.

Phase 2: Plan the Project

The Uncertainty Continuum of Risk

C	B	A
0% Certainty 100% Uncertainty	Certainty > 0 < 100 ⟵⟶	100% Certainty 0% Uncertainty
No Information	Some Information	Complete Information
"We Don't Know What We Don't Know"	"We Know What We Don't Know"	"We Know"
The Unknown Unknowns (Unk Unks)	Traditional Domain of Risk Management	"Issues"

This diagram shows a common depiction of risk as a continuum and is a good way to explain the difference between *risks* and *issues*. It is also a good way to show that there are real risks afoot even if you and your team have not been able to identify them.

Segment A (right column) shows items that have 100 percent certainty of happening. Obviously, these are no longer risks, but they are issues and should be managed separately from your Risk Management Plan.

Segment B shows risks that have less than 100 percent probability of happening and items that have more than zero. This segment is what is generally called "risk management."

Segment C shows risks that exists, you and your team know they exist, but you are not sure in what ways they exist. This area is generally called the "Unk Unks," which stands for Unknown Unknowns.

Alternatives You Can Use to Manage Risk

You must deal with risk. It is part and parcel of managing projects ,and integral to the reality of your organization. The one thing you cannot do is to pretend it is not real. Your options:

1. **PREVENT IT:** Take steps to ensure that the risk or the consequence of the risk is avoided. The most desirable option is to ensure that the risk does not happen in the first place.
2. **MITIGATE IT:** Here is where corrective and remedial action come into play. Part of your mitigation plan is to know what you will do if it does happen. Part of mitigation might include revising the scope of your project based on the business objectives of your organization. For example, you might reduce the scope in order to lower the risk to a more acceptable level. These are difficult trade-offs and must be made in the context of your business objectives. Sometimes you may have to take the risk in order to achieve your business objectives. The ultimate mitigation step is to eliminate the root cause of the risk. Sometimes we use the band-aid approach, where we treat symptoms of the risk but not the actual root cause. Methodologies such as the Fishbone Diagram, FMEA (Failure Mode and Effect Analysis) and TRIZ lend themselves well to root-cause identification.
3. **TRANSFER IT:** Move the risk to another entity; for example, outside consultants or contractors.
4. **DELAY IT**: Choose to delay the risk by postponing the trigger as much as possible.
5. **ACCEPT IT**: Make it part of your business model and choose to tolerate the consequences; that is, take the bullet. This may be your best option if the expected loss is lower than the other risks or it will be simpler to absorb the impact than to deal with it.

Pretending that the risk does not exist or ignoring it is not an option. . Sooner or later, risk will impact your project. And this is the head-in-the-sand approach is simply a form of denial. The smart project manager ensures that in spite of denial, the risk is defined and mitigated.

Example: Focussed Risk Management

	Level of the Risk Management Plan	
Types	**Program Level**	**Project Level**
Financial (Business Case)		
Marketing (The Market Place)	Marketing will develop and sustain a risk management plan whose focus is the competitive environment. The objective is to anticipate and mitigate the actions of our competitors.	
Engineering Design (Software and Hardware)		Design engineering will develop a FMEA focused on the new cooling system. The objective is to mitigate any risks due to the new design and new cooling requirements.
Manufacturing (Process)		Manufacturing will develop a Process FMEA to address the new assembly requirements of the new cooling systems.
Supply Chain (Suppliers)	The program manager will develop and sustain a risk management plan to ensure that our suppliers from Asia, India and Belarus can deliver on-time and on budget.	
Service and Repair (Field service, repair and maintenance)		

Risk management can be done at many levels, it does not have to be done at the project level. Your project may need focus on risk on a specific area, where there is more leverage for you and your team to make a difference. For example, the above matrix shows that you can do a risk management effort only on the supply chain due to the nature of your project. Likewise, you could do a program level risk management for the service and repair associated with your new product.

In large product development there can be several, if not many, projects; in that case, Program Management comes into play; i.e., the management of several projects with a common outcome. Incidently, there can be several risk management plans in a single program.

Use the above matrix to collaborate with stakeholders to define the type and area of risk management for your project. There are two axes in the above matrix. The first column shows the specific function in your organization, in this case the one which is responsible for generating the risk management plan. The second shows the program and project level.

Phase 2: Plan the Project

Types of Risk Management to be Performed

Risk management can be performed at various levels, depending on your objectives, although it is imperative that your team has full clarity as to the type and level being applied. For example, you can do a Risk Management Plan at the marketing level that could address the risks of introducing a new product or the risks with entering a new market.

You can create a risk plan with emphasis on your supply chain. At the product level, you can start with a component-level risk plan, then a component-level and finally a full system-level plan. Again, the plan will be driven by your objectives and judgment.

Finally, there can be several Risk Management Plans associated with a single program. For example, marketing, engineering and manufacturing can work on three plans concurrently. In this situation, it is necessary to provide coordination and cross-functional support to avoid duplication and take advantage of one another's findings.

Notice the example in the following page. It defines the types of risk management to be performed for a particular development program. Naturally, because every program is different the type of risk management will also be different. You can use the template in the following page to help you map your risk management strategy.

Essential

As part of this series, we have developed a guidebook that deals exclusively with the risk management in product development. Email the authors for additional information.

See Also

Risk Management and FMEA for Product Development

Example: Types of Risk Management to be Performed

In the hypothetical example below, there is a combination of Risk Management Plans due to the complexity of the program. Notice that there are more than technical risks to this program; consequently, the need to address other areas is very important.

	Level of the Risk Management Plan			
Types	**Program Level**	**System Level**	**Subsystem Level (Module)**	**Component Level**
Financial (Business Case)	Product management will develop, document and present for approval a Risk Management Plan on the financial risks of the new product, vis a vis the assumptions in the Product Proposal.			
Marketing (The Market Place)	Marketing will not be required to develop a plan for this program.			
Engineering Design (Software and Hardware)				Engineering will develop, document and present for approval a RMP for all new components.
Manufacturing (Process)		Manufacturing will develop, document, and present for approval a Risk Management Plan for the new assembly process and for outgoing inspection.		
Supply Chain (Suppliers)			Supply chain will develop a Risk Management Plan regarding all our suppliers in the Far East.	
Service and Repair (Field service, repair and maintenance)			Service will develop, document and present for approval a RMP for depot-level service.	

Phase 2: Plan the Project

Risk Management

Overview

The process to develop and implement a Risk Management Plan can be reduced to three phases, as shown in the graphic to the right. This view provides a simple, yet realistic view of risk management. Additionally, we use this graphic to help you navigate through the phases as they are outlined in this guidebook.

Note that some suggest that there are only two phases, followed by implementation of the plan. We acknowledge that this is also valid, and the end is the same; that is, managed risks.

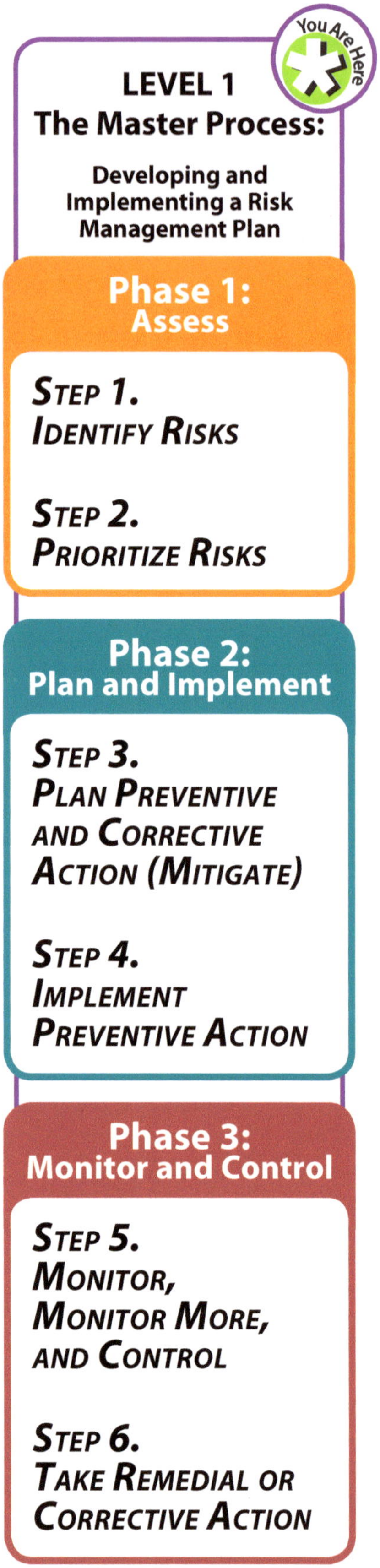

Phase 1: Assess

Identify and prioritize the risks. This is where your team has the most participation and you lay the foundation for all subsequent steps. Most risk management methods rely on a **QUALITATIVE** approach; that is, they rely on the experience and collaboration of your team to identify risks. **QUANTITATIVE** risk management falls under the category of "reliability engineering," which is not covered in this guidebook.

Phase 2: Plan

Once the risks have been prioritized, you assign the mitigation for each risk. Mitigation includes preventive and corrective action. In plain English, it answers two key questions: What can we do to prevent this risk from happening? And what are we going to do if the risk does happen?

This phase also includes the documentation of your plan such that it is available to your entire team.

Phase 3: Monitor and Control

This phase is the ongoing effort to monitor risks and update new risks. Here is where the concept of proactive risk management comes into play. It is not enough to develop a plan and file it. You must continually monitor risks and add new risks as they are identified. A schedule of regular meetings is the best way to implement the proactive risk management concept. Wise organizations schedule meetings for the sole purpose of monitoring and updating the risk plan; this way you do not shortchange your ability to manage risk.

Obviously, this phase is where you put your plan into action should any of the risks need mitigation.

What You Will Learn

In this guidebook we provide a fully documented and detailed process of each of the steps in the three phases.

First, you will discover the detailed process map for risk management. Following that, the risk description and risk prioritization are explained. Also see Critical Tools at the end of this guidebook for additional information.

Phase 2: Plan the Project

Before You Start

Checklist

- ❑ **CALL A MEETING:** Your objective is to call a meeting with the participants in your risk management efforts. This meeting is the first tangible manifestation of your efforts; consequently, you must prepare to ensure a successful event.
  - This first meeting is also a brainstorming session. You want to invite stakeholders plus your risk management team (if anyone has been selected). The intent of this meeting is to get as much input as possible from as many stakeholders as possible.
  - Also note that this first meeting is bound to be long, from four hours to as many as eight. Your guests should be aware and agree to be at the meeting for the duration. It could also be two sessions of four hours each for additional information).
- ❑ **COORDINATE LOGISTICS:** For example, schedule a venue that is large enough for your team to work, and make sure you have the right supplies. What you will need:
  - A large conference room to allow for much interaction.
  - An ample supply of yellow stickies, as it is the principal tool for the task.
  - One or, preferably, two flip charts with fresh paper pads.
  - Markers.
  - Masking tape.
- ❑ **SELECT ATTENDEES:** Select the names of attendees on your risk management team who will attend the meeting; do this based on their ability to help contribute to risk management planning. Ensure diversity of participants — the more points of view the better. For example, combine people who are involved in your project with people who aren't. People from other functions may be helpful, as would some of your suppliers and vendors. Use your imagination to ensure diversity. Who to invite? For example:
  - The members of your team (Marketing, R&D, Service, Quality, Manufacturing, Etc.).
  - Key stakeholders.
  - Key managers or supervisors.
  - It may make sense to invite key suppliers, contractors and consultants. They play a critical role in the success of your development project.
- ❑ **PREREQUISITE TASKS:** Prior to calling this meeting you should have accomplished the following:
  - Completed the Team Charter Document.
  - Selected your Risk Methods and Strategies using the Complexity Diagram.
  - Completed the Types of Risk Management to be Performed template.
  - Already have a good idea of the members of the Risk Management Team, if you chose to have one.
  - Already have the name and title of the Management Mentor (Sponsor), if you chose to have one.

Phase 2: Plan the Project

Important Notes

Try to keep the group to fewer than 10; more participants adds a level of complexity that may make it difficult to succeed. If at all possible, retain a facilitator. This will expedite the process and make it more productive. Some companies have internal facilitators; your HR department may have a qualified person. You may wish to ask a manager or another project leader to act as facilitator.

Avoid the creation of a bureaucratic document; rather, the Risk Management Plan is a dynamic tool that helps you monitor and take the appropriate action. We encourage flexibility in the definition of the format of your plan. In this Guide we provide two approaches and several examples to allow you to customize to your and your team's specific needs.

This step-by-step process will guide you through the development and implementation of a risk management plan. It may take several calendar weeks to complete the entire process, and that's acceptable. Also, you may need to make minor changes to the process to adapt it to your business objectives and the culture of your organization.

Once you have made the decision to have a risk management plan, the first step is to make all the preparations for the meeting. This is the first active result of your efforts that is visible to the entire development team. By the time you are ready to call the meeting you may already have organized the risk management team, although it can be done as an outcome of the meeting.

Detailed Process: Risk Management

1

Planning and Staging

Being well-prepared will make risk planning much more effective. As with any project, planning is critical.

Once planning is complete, proceed to invite the participants to your first risk management meeting.

Important

Part of staging is the completion of the charter document.

Ensure the support of the proper managers and supervisors.

See Also

Teams and Team Charters

2

Invite the Participants to the Planning Meeting

This is your risk management meeting. Allow enough time for creativity, interaction and dialog. These meetings last from four hours up to a full day. Don't shortchange the process — the time spent in risk planning will pay off later.

Who should attend? Those people who can contribute to identifying anything that could go wrong with your project.

Important

Risk identification is a brainstorming exercise; allow for much interaction and ample time for dialog.

During your brainstorming, review previous projects to help you identify additional risks.

See Also

Type of Risk Management to be Performed

Continued

CONT'D

3

START THE MEETING

Start by stating the purpose of the meeting, which is to identify and prioritize risks for your project.

Present a quick overview of the project; for example, use the Gantt Chart to show the schedule. You can also show a Block Diagram, and any other information that may add clarity of understanding.

1. Present the Team Charter
2. Present the Risk Complexity Diagram
3. Present the Type of Risk Management to be Performed

Ensure that everyone has sufficient understanding of your project so they can better contribute.

Describe the project and its objectives, adding any information that might be helpful; for example, block diagrams, photographs, videos or flow charts.

See Also: **Protocol for Risk Definition**

4

INITIATE THE BRAINSTORMING TO IDENTIFY POTENTIAL RISKS

1. Ask the participants: "What are all the things that could go wrong with this project?"
2. Ask everyone to write their ideas on yellow stickies. It is best to request three ideas per person to start, one idea per yellow sticky. By now they should be familiar with the Yellow Sticky Protocol, but if not go over the protocol before they start writing.
3. After a reasonable period of time, ask each participant one at a time to read aloud his yellow stickies and place them on the board or flip chart.
4. After everyone has read their three ideas, ask for more ideas. Encourage people to write down anything and everything that could go wrong with the project. Every new yellow sticky will create new ideas, and this is part of the process. Continue to encourage your team members to create more risk possibilities. Reassure them that there are no bad ideas, and that the rest of the process will ensure objectivity.

Keep brainstorming until it is obvious that all ideas have been captured.

Time Frame: The process of identifying possible risks could take two to four hours. Don't rush it! This is the most important step in the process. People must have time to brainstorm.

5

CONTINUE THE BRAINSTORMING

At this point you should have many yellow stickies on the board or flip chart, perhaps 50, 100 or more.

Make sure they are all readable and understandable. If you need to change a yellow sticky, you must ask the author for permission and clarification. This is very important, as it may generate new ideas and add much clarity.

Important: Once you've finished brainstorming and all the yellow stickies are on the board, take a coffee break! Let your team rest in preparation for the next step.

CONTINUED

Phase 2: Plan the Project

CONT'D

6

CLEAN UP YOUR BRAINSTORMING

Start by discarding the obviously unlikely risks; for example, "We get hit by an asteroid." But you must ask the author for permission and clarification — this is very important.

Analyze by:

- Reviewing the description: Is it clear? Does the team understand the risk in the same way?
- Should there be criteria associated with the risk?
- Are there degrees to the risk?

See Also — Karnaugh Maps

Use the 2x2 Matrix to prioritize 30 or fewer identified risks.

Important

You have two ways to prioritize. For a relative few number of yellow stickies — 30 or fewer — we recommend the 2x2 Matrix for prioritizing results (see Karnaugh Maps).

When you have a larger number of yellow stickies, we recommend the PI Factor approach, where P is the probability and I is the impact. The following steps take the PI Factor approach to prioritizing results.

7

PRIORITIZE BY PROBABILITY

Now prioritize the remaining risks by their likelihood of actually occurring.

You can use a scale from 1 to 10, where 10 is the highest probability. Work with your team to determine the value for each of the yellow stickies. Write the value for the probability in one of the corners of the yellow sticky.

You can do a second round of prioritizing, this one using a more scientific analysis of the priority; for example, analysis based on a percentage, where each level is defined using documented criteria. This step is not necessary every time, but only when additional accuracy is needed.

NOTE: This is one of the two priority-setting steps.

See Also — Developing Scales for Risk Management

8

PRIORITIZE BY IMPACT

At this point, you have prioritized by likelihood — i.e., the probability of the risk happening.

Next, go over all the prioritized ideas, but this time prioritize by asking, "What will be the damage if it does happen?" That is, how severe will the impact be if it does happen?

As you did in the prior step, assign a number from 1 to 10, where 10 is the most severe impact and write the value for the impact in one of the corners of the yellow sticky

Identify the few that have a high probability of happening and at the same time a high level of damage if they do happen. These are the "vital few."

9

CREATE THE RISK FACTOR

By now, each yellow sticky should have two numbers. One for the Impact and another for the Probability.

To create the Risk Factor simply multiply the two numbers. Write the product on the yellow sticky.

Once you've calculated and recorded the Risk Factor for every yellow sticky, you can enter this data into a spreadsheet.

CONTINUED

Phase 2: Plan the Project

Cont'd

Where You're At

At this point you have:

- Identified the risks.
- Prioritized them by probability.
- Prioritized them by impact.
- Created a PI Factor for each yellow sticky.

Important

The impact can be measured in several ways — your choice. The most common way is to translate into dollars. The impact of every risk can always be translated into the language of dollars. Another choice is "lost days" or "slippage" — the number of days your project will be delayed if the risk happens. Another choice is the impact on the customer — i.e., the negative impact on customer satisfaction.

10

Document Key Assumptions

In many cases, risks are predicated on something else happening. That is, the risk will occur only if something else takes place. These are called key assumptions. For example, "Our budget gets cut" leads to the assumption that the damage or impacts on your project are predicated on the budget being cut. Another example is "There is a company-wide layoff." In this case, the damage to your project is predicated on the assumption that it will occur "if" there is a layoff.

Document all these assumptions. They will be very important in managing risk as your project progresses.

Important

It may be a good idea to end the meeting at this point. If you detect that your team members are fatigued, it is best to schedule a second meeting to define preventive and corrective actions.

Where You're At

At this point you have prioritized each risk based on Probability and Impact. Each yellow sticky will have three numbers written on it: the Probability, the Impact, and the Risk Factor, which is the product of the Probability and Impact. The Risk Factor provides the priority. Naturally the higher the Risk Factor, the more important the risk is.

11

Start Development of the Risk Management Plan

It's essentially impossible to create a Risk Management Plan for every yellow sticky, particularly when you have over 100, so you and your team need to decide how many of the prioritized yellow stickies you will address as part of the plan.

You have the Risk Factor for each; naturally, those with the highest Risk Factor get the priority. You can use your documentation (Charter, Complexity Diagram, and Types of Risk to be Performed) to help you and your team decide.

Once you've prioritized and decided how many you will address in your plan, it's time to start defining mitigation — your Risk Management Plan. For each of the risks you identified as being high probability and high impact, develop a documented action plan for preventive measures and corrective actions.

Continued

Phase 2: Plan the Project

CONT'D

12

DEFINE YOUR MITIGATION ACTION

PREVENTIVE ACTIONS are those activities you will undertake to prevent the risk from happening — i.e., you anticipate and thus prevent.

CORRECTIVE ACTIONS are those activities you will undertake if the risk does take place.

For each of the risks you identified as being high probability and high impact, develop a documented action plan for preventive measures and corrective actions.

Once your plan is complete, review with the project sponsor, your manager and those key stakeholders you deem appropriate.

Distribute copies of the plan to your team members.

Archive a copy in a safe place. In some cases, this is a required step due to regular formal meetings throughout.

Important!

The most valuable benefit from risk management is discovering the steps you can take, right now, to ensure that the risk never happens.

Pay particular attention to preventive action — you will have the highest return on you investment by making certain that the risk stops being risk. Ensure, right now, that it will never happen!

WHERE YOU'RE AT

At this point you have:

- Defined preventive and corrective action for each prioritized risk.
- Documented key assumptions (via yellow stickies and flip charts).

The meeting ends when the above list has been completed.

13

DOCUMENT YOUR RISK MANAGEMENT PLAN

You now proceed to document your plan. There are several templates in this guidebook you can use. It's a good idea to solicit the help of a few members of the team to help in the documentation — it serves to create ownership, plus it adds clarity to your document.

The document can be an Excel spreadsheet, provided it's easy for all members of your team to understand.

14

DEPLOY PREVENTIVE ACTION

Once you have a documented plan of preventive actions, lead your team in implementing these actions. Also implement new preventive actions discovered after the document plan was drafted.

Here is where your risk management efforts start to yield results — you save time and aggravation, and you help achieve your desired results.

CONTINUED

Cont'd

15

Monitor and Report

Constantly review the list of prioritized risks and assess the need for preventive action. This is done formally and informally.

Formal review includes regular meetings with the Risk Management Team, the engineering team and other key stakeholders. The formal meetings should happen regularly. Use your judgment to decide the space between meeting; that is, weekly, every four weeks or something else.

Informal review occurs through direct interaction with all stakeholders.

Continually review your plan to derive its benefits. Do this through your regular risk management meetings.

Keep all stakeholders informed about risk.

Change and communicate the key assumptions as your project progresses. Changes to the assumptions also mean updating your plan.

The preventive and corrective actions may need to be changed as your project progresses.

Your agenda for your review meetings:

1. Review your plan.
2. Discuss variances.
3. Mitigate risks as needed.
4. Terminate risks that have been mitigated.
5. Identify and prioritize new risks.
6. Update your plan and create new plans for newly identified risks.
7. Publish your plan.

Important

Remember the Five Questions of Proactive Project Management

1. What is the variance between the baseline and the actual (schedule, budget, etc.)?
2. What are the issues causing the variance?
3. What is the current trend; that is, the preliminary predicted status if we continue with our current performance?
4. What new risks may pop up in the future, and how could they change the preliminary predicted status?
5. What actions should we take to prevent the predicted state from happening and deliver on the baseline?

16

Take Action

YOUR PLAN IS OF NO USE IF YOU DO NOT ENSURE ACTION — A.K.A. IMPLEMENTATION!

- Take preventive action as soon as you complete the plan.
- Take corrective action when a risk cannot be avoided.
- Continue to hold your risk management meetings to monitor and mitigate risks
- Update your plan as needed.

Learn and Improve (Kaizen)

Your risk management process will improve significantly if you constantly add the lessons you've learned.

- After your project is completed, identify how well your Risk Management Plan worked.
- Interview team members and capture their suggestions for improvement.
- Learn from the experience — document and use lessons learned.
- Build Best Known Practices (BKPs) for risk management so that you and others will have the information available for your next project.

Phase 2: Plan the Project

Protocol for Risk Definition

Introduction

Every risk management methodology starts with the identification of risks; this is intended to capture everything that might be remotely considered a risk. It is a brainstorming phase that relies on the knowledge of a group of people.

Some organizations, based on experience, have a checklist for risks. These lists are compiled over time and based on similar projects. There is a danger (risk) with this approach: No two projects are the same, but a checklist may create a feeling of completeness; that is, you might mistakenly assume the list you use is perfect and complete. The use of a checklist is acceptable, but keep in mind that new risks appear with every new project.

Part of risk identification may include research, either prior to the meeting or afterward. Research includes reviewing prior projects and risk management plans, and also interviewing people in your organization who participated on previous projects or have a special knowledge about a topic of interest. Read papers and articles on the topics of interest, do a Web search and even look into purchasing books on the topic. You can assign research to various members of your team and ask them to come prepared to make a presentation during your meeting.

Speaking of Risk

It is important that you and your team have clarity about the language you will use in defining risks. First, there is the risk itself, "causes the mechanical design to be delayed by five days." Then there is the "cause" of a risk or, as some call it, the "root cause"; for example, "the chief mechanical engineer calls in sick." Think of this as the source of the risk. Finally, there is the impact or effect of the risk, "resulting in a delay of three days for the entire project."

Three Parts

Thus risk, for the purposes of risk management, must be stated in three parts:

1. **RISK:** The actual risk.
2. **CAUSE:** The cause or root cause or source.
3. **IMPACT:** The consequence or effect.

For example:

1. **RISK:** "Competitor A confuses the market."
2. **CAUSE (SOURCE):** "Competitor A introduces a new technology."
3. **IMPACT (CONSEQUENCE):** "Our sales are reduced by 25 percent in the first six months."

LEVEL 1
The Master Process:
Developing and Implementing a Risk Management Plan

Phase 1: Assess

STEP 1. IDENTIFY RISKS (You Are Here)

STEP 2. PRIORITIZE RISKS

Phase 2: Plan and Implement

STEP 3. PLAN PREVENTIVE AND CORRECTIVE ACTION (MITIGATE)

STEP 4. IMPLEMENT PREVENTIVE ACTION

Phase 3: Monitor and Control

STEP 5. MONITOR, MONITOR MORE, AND CONTROL

STEP 6. TAKE REMEDIAL OR CORRECTIVE ACTION

Three Parts Defined Another Way

1. **DESCRIPTION OF THE RISK:** "The risk that we run out of fuel during our trip." Notice that this part only says what the risk is; it does not go beyond a simple description. Also note that it is short, but easy to understand. We recommend you avoid cryptic descriptions — e.g., "Run out of fuel."
2. **POSSIBLE ROOT CAUSE:** As you state a risk you are also making mental assumptions about the cause of the problem. It is not required that you state the root cause, but it is highly recommended.
3. **IMPACT:** "It causes us to lose $500 due to late fees." This part describes the actual damage or consequence as a result of the risk actually happening.

Every risk must have at least one impact (consequence). A single risk can have several impacts, each calling for different mitigation. Stating the actual impact for every risk ensures completeness; that is, avoid the trap of assumptions, where one person believes that his perspective of a risk has been stated by someone else.

It is important that there is agreement within the team as to how the risks are stated to ensure clarity and convenience. Everyone on your team must be able to easily understand the meaning of the risk as it is described. Remember also that others in your organization and beyond (e.g., vendors and suppliers) may have to deal with the risk. Enforce the three-part description, but accept occasional exceptions.

Using the Yellow Sticky Protocol to Document Risk

The need for a protocol, or set of rules, to describe a risk is necessary to ensure communication among the team members and with others not in the team but with whom you will need to collaborate to deal with the risk. A protocol saves time during the development, as everyone on the team knows what they are reading, and in what order.

The ability to ensure that everyone in a group participates is greatly enhanced through the use of Post-It Notes® (trademark of 3M), known in the vernacular as "yellow stickies." A benefit of the yellow stickies is that they provide a self-documenting process. Also, yellow stickies significantly accelerate the documentation, rather than writing things on a flip chart, which is time-consuming. Finally, yellow stickies are portable; that is, one can move them, hand them out, rewrite them, etc.

The risk management process used in this guidebook is **QUALITATIVE** in nature, which means the actual risks will be identified by your team members during the risk management session. As such, there will be a variety of ways that your team members will use, from cryptic "the pump" to very detailed "the two-stage back up pump will fail if the pressure exceeds 200 PSI." Incidentally, for **QUANTITATIVE** risk one needs to look more to the area of reliability engineering, which is not part of this document.

As we recommend in the process map, all risks will be written in yellow stickies; furthermore, we strongly suggest that you create and enforce a protocol for the way the risks are described. This will ensure clarity and uniformity that will save you time and aggravation.

In the following pages we show examples and a description of a protocol. You need not use the approach we describe, but you need to use a protocol for the sake of clarity and efficiency. It is critical that every yellow sticky has the following information: the description of the risk, the possible cause, the impact if the risk were to happen and the name of the person who wrote the note. Feel free to require additional information if it helps you with clarity and efficiency.

Important! In some organizations this protocol is known as the "if / then" method — if such and such happens, then such and such takes place.

Phase 2: Plan the Project

Examples: Using Yellow Sticky to Document Risk

Example 1: The Protocol in General

Shown here is the Yellow Sticky Protocol for documenting a risk. Notice the description shown in the various call outs. Feel free to modify your protocol to add clarity and efficiency. The intents are to clearly describe the risk and its impact and to trace the yellow sticky to its source.

Example 1

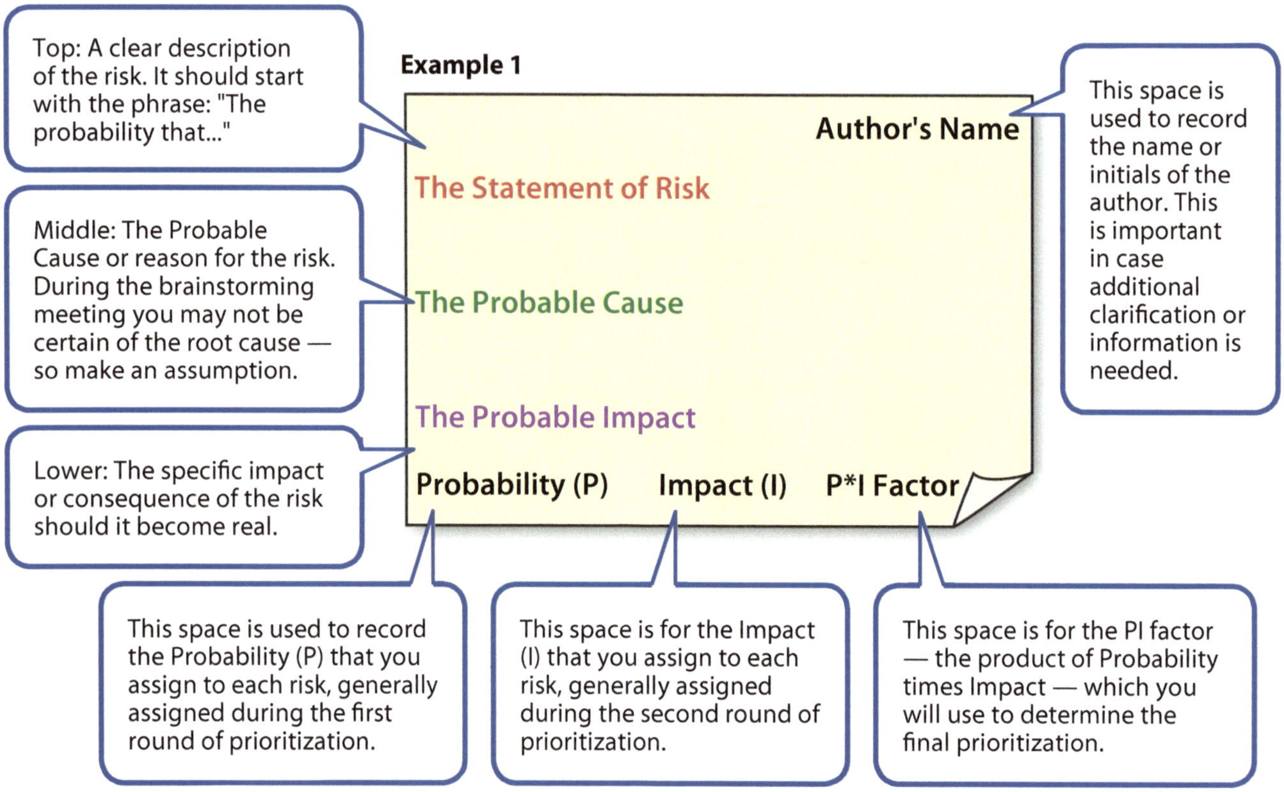

Example 2: The Incorrect Protocol

This is poor way to describe a risk. Note that the information is vague, it has no context; therefore, it can be interpreted in many different ways. You don't know who wrote it; consequently, it will be difficult to get a clarification. When someone writes a risk with too little information, help him rewrite it using the protocol.

Example 2

For comments, questions or additional copies please e-mail the author: flexpm@rapidinnovation.com.

Example 3: The Correct Protocol

These examples illustrate the correct way to document a risk during your brainstorming session. The presence of the author's name allows for the opportunity to clarify. Also note that these examples have already been prioritized and include the PI Factor, which will enable the final prioritization.

Example 3-A

José C.

The probability that the new user interface does not meet customer requirements due to lack of usability research consequently making the new product difficult to use

P = 6 I = 8 PI = 48

Risk Statement

Probable Cause

Impact of Risk

Example 3-B

Mary M.

The probability that the new power supply will not be able to sustain the environmental requirements due to the low environmental ratings results in overheating causing catastrophic failure and potential for fire

P = 6 I = 8 PI = 48

Risk Statement

Probable Cause

Impact of Risk

Example 3-C

Jan J.

The probability that the main water pump will develop cracks due to the high temperature of the water, causing the cooling system to shut down, which would stop production

P = 6 I = 8 PI = 48

Risk Statement

Probable Cause

Impact of Risk

Phase 2: Plan the Project

Example 4: Impact or Consequence?

In risk management, we always talk about the impact that a particular risk will have on your project. Impact is the negative effect that a risk has on the successful completion of your project. Consequence is another word for impact. Use either one. In FMEA, the preferred term is "impact."

Notice that "due to poor financing" is the possible root cause.

Example 4

The probability that our main supplier in Singapore goes out of business due to poor financing.

Causes a disruption in our JIT process, thus shutting down our manufacturing plant for three weeks.

First session Bill A.

Example 5: One Risk with Two or More Possible Impacts

In some instances, one risk can have two or more possible impacts. This is yet another reason for clarity in describing risks. It becomes essential; for example, consider the ramifications of two team members talking about the same risk, but having two different impacts in mind. Everyone must describe the risk using the protocol, to ensure absolute clarity.

Example 5-A

The risk that our software is vulnerable to security attacks.

Causes us to fail the acceptance test with our main customer, and thus the loss of a very large order.

Session in Germany Bill A.

Example 5-B

The risk that our software is vulnerable to security attacks.

Causes a major change in architecture, which results in a six-month delay to our rollout plan, and the loss of $1,000,000 in sales.

Session in France Bill A.

Example 6: The Wording of Risks and Potential Failures

The way a risk is stated is very important for the sake of clarity and also for processing the information. Below are additional examples that might help you articulate your risks.

Example 6-A

Notice that "due to the brittleness" is the possible root cause.

Bending the power cord during manufacturing causes it to become frayed due to the brittleness of the sheath.

Results in a safety risk that might produce electric shock to the point of severe injury to the customer.

Session "A"; Group "2" Bill A.

Notice that in every case the impact is worded from the customer's point of view.

Example 6-B

It is very likely that the radiator will leak after 12 hours of use, which would cause the engine to overheat and trip the temperature control shut-down relay.

Results in making the vehicle inoperable for the customer.

Session on the engine block Bill A.

In this example, the root cause is not stated; that is, the reason that the radiator leaks is not identified.

This is acceptable if the root cause will be addressed later in the process.

Notice that the code to trace the information is the risk management session.

Example 6-C

There is a probability that the operator opens the safety cover and touches the high-voltage supply.

Results in the death of the operator.

020507, 10AM Bill A.

This is clearly a show-stopper risk. Safety is always of extreme importance.

Phase 2: Plan the Project

Example 7: A Chain of Risks

In some instances, one risk points the way for another risk. This is called a "chain of risks." This is a very important topic. Studies have shown that teams tend to overlook the risks caused by another risk-taking place. We suggest that you probe for additional risks that might be triggered by an initial risk.

Example 7-A

It is very likely that the radiator will leak after 12 hours of use, which would cause the engine to overheat and trip the temperature control shut-down relay.

Results in making the vehicle inoperable for the customer.

Engineer, EE — Bill A.

Example 7-B

Overheating the engine causes failure of the water pump due to the high temperatures of the water.

Results in making the vehicle inoperable for the customer.

Engineer, ME — Mary J.

Example 7-C

Failure of the water pump causes the engine block to crack.

Results in a very expensive repair bill for the customer.

Technician — Bob D.

Prioritizing Risks: Creating a List of the Vital Few

Introduction

Once you have identified all possible risks, it's time to prioritize them based on solid criteria that you and your team create. Prioritization can be achieved through the use of Affinity Diagrams. There is a full description and process map for Affinity Diagrams in the Critical Tools section of this guidebook. An additional method to prioritize is through a Karnaugh Map, which also is explained in the Critical Tools section of this guidebook.

The strategy for prioritizing is to enable your team to focus on the vital few; that is, those risks that are the most likely to occur in your project and can also have the most impact. The examples on the following pages show the mapping of risks using a Karnaugh Map.

Simply stated, not all risks have the same impact. Some can be very minor and insignificant, while others can indeed inflict major impacts on your project. By the same token, not all risks have the same probability of happening. Some are so remote that it's not worth worrying about them, while others have a high probability of happening. Consequently, it stands to reason that you want to sort risks by the highest probability of happening and by the highest impact. This combination provides the list of the vital few.

Once you have prioritized your risks, you need to develop your Risk Management Plan.

Karnaugh Maps (2x2 Matrixes)

Facilitators are often faced with helping a team make a decision or set priorities from a selection of many options. Generally, when choosing, we have two criteria — for example, Probable and Important. While some things can be very important, they are not necessarily probable. Thus, the ability of a team to separate all tasks based on the criteria will result in a better deployment of limited resources.

A Karnaugh Map is a participative tool that enables a group to prioritize or select items from a group. A Karnaugh Map is a 2x2 matrix that highlights how the criteria relate to an item and enables better and faster decisions to be made.

A PI matrix specifically highlights the probability and importance of a risk.

See Also: Karnaugh Maps

See Also: Risk Management Plan

LEVEL 1
The Master Process:
Developing and Implementing a Risk Management Plan

Phase 1: Assess
Step 1. Identify Risks
Step 2. Prioritize Risks — You Are Here

Phase 2: Plan and Implement
Step 3. Plan Preventive and Corrective Action (Mitigate)
Step 4. Implement Preventive Action

Phase 3: Monitor and Control
Step 5. Monitor, Monitor More, and Control
Step 6. Take Remedial or Corrective Action

Phase 2: Plan the Project

Examples: PI Matrixes

Each of the four corners represent the maximum value for at least one of the axes. In this case, the maximum (10) refers to the probability of the risk happening.

When the scales are the same (0 to 10), then this quadrant contains the most important risks; i.e., those with very high probability of happening and the most damage if they were to happen.

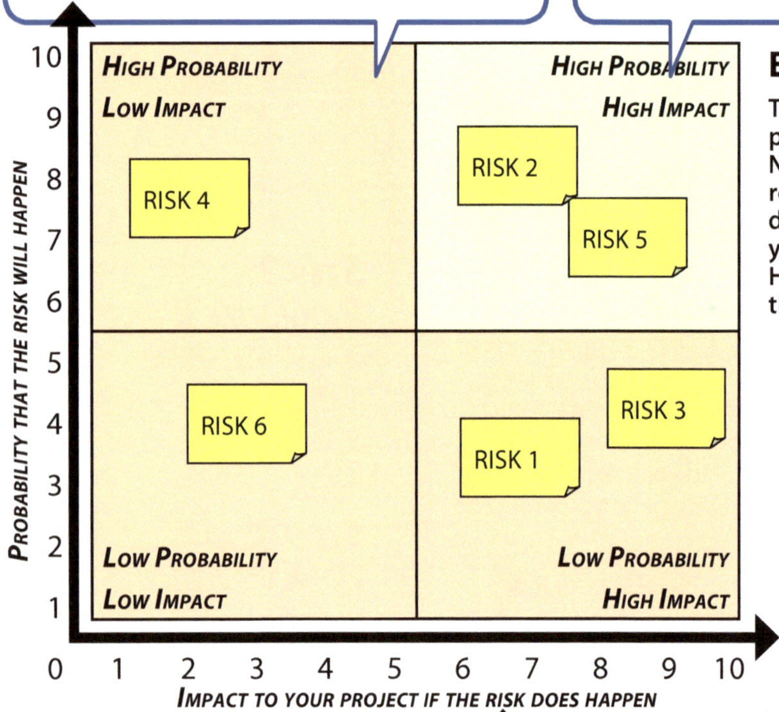

Example 1

The y axis shows a scale for the probability of a risk happening. Notice the 0 to 10 scale, which is the recommended approach. The x axis depicts the impact; i.e., the damage to your product if the risk were to happen. Here the scale is also 0 to 10, 10 being the maximum impact.

This quadrant is the most important, because the risks are likely to happen, and there will be a serious impact if they do happen.

Example 2

This example demonstrates a team effort to prioritize risks associated with building a new house. Notice the upper right quadrant where the most important risks are mapped — those risks have the highest possible impact and are most likely to happen.

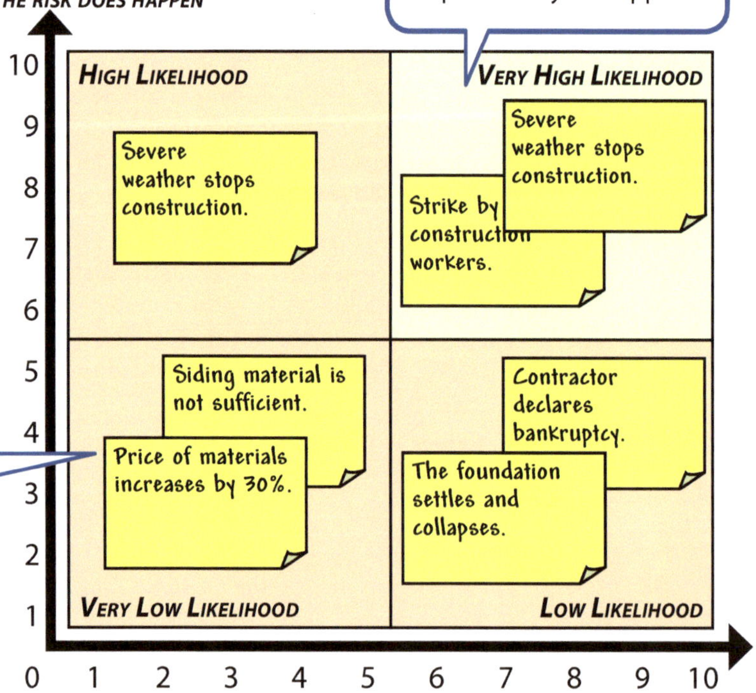

This is the "don't care" quadrant. Items in this area are not likely to happen, and even if they do, the impact will be miniscule.

Example: Tabular PI Matrix — Enter Data into a Spreadsheet

In the previous section you discovered how to use Karnaugh Maps as a way to prioritize risks. In this section you will learn how to apply the PI Matrix results to a spreadsheet. Both methods are acceptable. Depending on your objectives and the style of your team, select whichever fits best.

After the risk identification session, create a row for each risk and identify the risk in the Item column. Ask your team to rate each risk.

First assign a rank from 1 to 10 for the Probability. The higher the number, the more likely that the risk will happen. Enter the number into the Probability column.

Next ask the team to assign a number to each risk signifying its Impact. The higher the number, the more negative impact the risk will have. Enter the number into the Impact column.

Determine the PI factor, which is the Probability ranking times the Impact ranking. Once you have entered the data into your table or spreadsheet, it's a simple matter to sort the data by the PI factor. You can use this spreadsheet to develop your risk management plan.

Example: Enter Data into a Spreadsheet

Item	Probability	Impact	PI Factor	General Notes
1. The pump fails under pressure.	6	8	48	Assumes constant temperature.
2. The filter allows foreign matter.	4	2	8	

In the above example we have listed only two risks for the sake of simplicity. Note that we have entered a number between 1 and 10 to represent the Probability that a risk could happen, and also the same scale for Impact to indicate the severity of the damage if the risk were to happen. The multiplication of both numbers (Probability and Impact) will yield the PI factor. When executing the sort command in an Excel spreadsheet, you can use the PI factor to prioritize risks. You then need to decide how far down the list you will choose to manage — i.e., the number of risks for which you will develop a plan, and continually monitor.

Definition

PI FACTOR

The probability ranking times the impact ranking.

Phase 2: Plan the Project

The Risk Management Plan

Definition

The mission of the Risk Management Plan (also called the Risk Response Plan) is to document and communicate risks, identify actions to reduce threats to the goals of the project and define clear responsibility for action.

The Risk Management Plan has at least four key sections:

1. **DESCRIBING THE RISKS IN ORDER OF PRIORITY** allows you to create the document of reference by which your team and managers can understand, identify and obtain the priority of risks for a specific project. It tells the reader which risks you have identified, which priority you have assigned and which risks you and your team have chosen not to track.
2. **PREVENTIVE ACTION** documents the specific steps you will take in anticipation to prevent the risk. In a best-case scenario, you can identify actions that will preclude the risk altogether.
3. **CORRECTIVE ACTION** documents the specific activities you will undertake should the risk occur.
4. **CLEAR RESPONSIBILITIES** documents who is responsible, when activities should happen and in what sequence.

What a Risk Management Plan Should Contain

There are certain elements that must be shown in order to define, track and manage risks. There are additional items contingent upon your circumstances. You must take into account the culture of your organization, the importance of the risk plan (every project is different) and the resources available to your team; for example, how many members and how easily can they meet. Use your judgment. Over time you will develop a Risk Management Plan that is unique to your organization and serves your objectives. Review the list on the next page to determine the elements you should include in your plan. There is flexibility; so, the following list is not rigid. The list simply reveals possible items in a Risk Management Plan.

In conclusion, your plan must contain enough information to enable your team to monitor and take action; in some companies the list of items in the plan is already defined. We suggest you review the list and ensure that it is still valid. Regardless, your plan must stand on its own, be clear and enable your team to implement it.

Essential

"Your plan must contain enough information to enable your team to monitor and take action."

Flexibility

The contents of your Risk Management Plan should reflect the culture of your organization, the objectives of your projects and your personal style. Be willing to experiment to find the best design for your company.

LEVEL 1
The Master Process:
Developing and Implementing a Risk Management Plan

Phase 1: Assess

STEP 1. IDENTIFY RISKS

STEP 2. PRIORITIZE RISKS

Phase 2: Plan and Implement

STEP 3. PLAN PREVENTIVE AND CORRECTIVE ACTION (MITIGATE) — You Are Here

STEP 4. IMPLEMENT PREVENTIVE ACTION

Phase 3: Monitor and Control

STEP 5. MONITOR, MONITOR MORE, AND CONTROL

STEP 6. TAKE REMEDIAL OR CORRECTIVE ACTION

Example: Possible Items in a Risk Management Plan

Item	Description
1. Work Package Number	In project management, it is the number of the specific task or work package.
2. Risk Event Number (Risk Number)	A sequential number assigned to every risk. Used simply for tracking purposes. In project management it can be the work package number, but you may choose to use a different sequence or format.
3. Risk Event (Risk)	The description in plain English of the risk at hand. For example, "Snowstorm prevents work.” See Also: Protocol for Risk Definition
4. Risk Probability	The likelihood that a risk will occur. Expressed as a number from 1 to 10, 1 being the least likely and 10 being the most likely.
5. Risk Impact	Severity of its effect on the project objective. Also called "risk consequences" or “amount at stake." This item describes the damage if the risk becomes real. Expressed as a number from 1 to 10 where 10 equals the highest impact; i.e., more damage if the risk happens.
6. Risk Event Status	Measure of importance of a risk event. Also called criterion value or its ranking. Similar to “criticality," a cumulative ranking of a risk based on probability and impact.
7. Project Risk	The cumulative effect of the chances of uncertain occurrences adversely affecting project objectives.
8. Criticality	The overall importance (priority) of the risk when compared to all other risks in the plan. This can be expressed as a number; this number is very important, as it is used to prioritize all your risks in relationship to one another. Also called Risk Factor, you can multiply the risk probability times the risk impact, and the product of the two is the risk factor.
9. Impacted Risk	When one risk affects another risk, either positively or negatively. Also called a “risk chain.”
10. Preventive	The actions you will take to prevent the risk from happening.
11. Trigger Point	The date or other signal that you will use to deploy the preventive step. Yet another action or step you will take if your “Preventive” is not available. Some call this “Plan B."
12. Owner	The person who will monitor and take action on the risk at hand. The individual who is accountable for managing a specific risk or risks.
13. Risk Event	The description of what might happen to the detriment of the project.
14. Contingency Reserve	The amount of money or time normally included into the project cost or schedule baseline to reduce the risk that the project objectives exceed an acceptable level.

Phase 2: Plan the Project

Item	Description
15. Management Reserve	The amount of money or time not included in the project cost or schedule baseline that is used by management to allow for unexpected and unpredictable situations.
16. Risk Financial Impact	A dollar estimate of the impact of a risk if it does happen. It can also signify the cost of preventive action. This is a very important number, as it assigns a meaningful figure (dollars) to your risk plan. Similar to Criticality but expressed in terms of financial impact.

This table contains a list of items, or topics, covered in a typical risk management plan. Your company may already have its own list, in which case you need to follow it. Not every item in the above list must appear in your plan. Rather, include every item that your team needs to monitor and take action on. You can collaborate with your team and decide the items, sequence and priority so it becomes a team list.

Risk Management Plan (RMP) as a Spreadsheet

Think of documenting your Risk Management Plan (RMP) in the form of a log using a spreadsheet. This format does away with excessive narrative and allows for more focused management of your plan.

Using a spreadsheet such as Excel is a convenient way to document your RMP. You can use the spreadsheet to capture, prioritize and monitor. Once you are ready, you can also document all the additional information you need to manage risks through the life of your development project.

Keep in mind that in a spreadsheet you may have several hundred risks captured, and when you are ready to document your plan in a spreadsheet, you have already identified, prioritized all the known risks using the PI factor (the product of the Probability times the Impact that you and your team designated to each of the identified risks). This happened in the brainstorming session you had with your team at the start of the process.

The PI Factor can be used in your spreadsheet to prioritize (for example, in Excel you can sort by the column that contains the PI Factor); this will rank each risk starting with those with the highest PI factor.

Once you have prioritized, you and your team need to decide how many of the risks you will manage. Use your judgment and remember that every development project is different; consequently, your choice will vary by project. For example in one you may chose to manage the top 25, while another may only require the top 10.

The information contained in a Risk Management Plan:

- The updates to the documented risks as more information becomes available.
- The dates where the review of the Risk Management Plan took place.
- New and emerging risks as information becomes available.
- Changes to the risk plan; for example responsibilities, course of action, etc.
- Documenting when a risk actually took place.
- Deletion of risks that will no longer happen or have become "issues."
- Other information that helps you track, and monitor your Plan.

Be Proactive

The Risk Management Plan is essential for proactive project management; it is the process of controlling risk. It calls for a high level of discipline and mental concentration. It calls for much initiative on the part of the project manager. It calls for an instigator, or activist, who is resourceful and takes action toward eliminating or mitigating the risk.

General Rules about the Risk Management Plan

1. Ensure that regular progress meetings are established to identify and mitigate risk. Ideally, this is a separate and independent meeting, which allows sufficient time and concentration to the critical task of managing risk.
2. Keep your Risk Management Plan up to date; while it may be a burden, it will save considerable grief and aggravation. Include the Risk Management Plan as part of your regular project review meetings or status meetings; this way you will be forced to keep it up to date.
3. Don't wait until the next regularly scheduled meeting to address a risk. Rather, call an urgent meeting, or even an informal meeting, to address the challenge.
4. Define the action needed to address the challenge. It is not enough to meet and talk about the risk. It is critical that you insist on identifying the action needed to address the risk. Always use the following checklist: **DO WHAT? BY WHOM? BY WHEN?**

Issue Versus Risk

Some organizations use the terms "issue" and "risk" interchangeably, but this is incorrect. The Issue Log tracks issues in the project, including current and future issues. Future issues constitute risk. That being said, the Issue Log and the Risk Management Plan can be combined into a single document and a single meeting.

- **RISK:** Something that has a probability, though not 100 percent, of happening.
- **ISSUE:** Something that has a 100 percent certainty of happening.

Therefore, an issue is something you must deal with, as it is an absolute certainty that it will happen. Stop thinking about it, and take action!

Essential

"An issue is something you must deal with, as it is an absolute certainty that it will happen."

Phase 2: Plan the Project

Documenting Your Risk Management Plan

In this guidebook we provide two approaches to documenting your risk management. Your choice is based on your objectives. We provide examples and templates of both viable approaches.

Approach 1: Traditional Table

The traditional method has been to use a simple table to document your plan. The advantages of this approach are the flexibility to send the plan around the world, the ability to update the plan as needed and the ability to insert considerable information directly into the document or as a hyperlink to other documents.

Example: Documenting Your Risk Management

Risk Management Plan

Work Package or Task	Risk Event No. and Description	Probability	Impact	Plan			Actual			Owner
				Risk Event Status (P x I)	Criticality	Impact on Other Risks	Preventive	Trigger Point	Contingent	
Develop blueprints	Draftperson not available due to illness	2	7	14	Critical	Yes Engineering	Hire another draftperson	Draftperson calls in sick	Borrow draftperson from another department	Fred Smith
Stone work completed	Stone mason from Europe unable to arrive on time	5	6	30	Critical	No	Coordinate travel arrangements well in advance	The stone mason is not on site 6 days prior to start	Hire local stone mason to start positioning the material	May Jones

Approach 2: Electronic Spreadsheet

A spreadsheet such as Microsoft Excel is another fine way to document your plan. The advantages of this approach are the ability to perform calculations, the sort command for instant ranking, plus the benefits of an electronic document that can be sent around the world and frequently reviewed and updated.

You can use the spreadsheet to capture and then prioritize risks. You can also document your management plan by entering the information next to each of the prioritized items in your spreadsheet. After quickly calculating the PI Factor, you can use then use Excel's Data Sort feature and instantly prioritize the risks. In a spreadsheet you may have several hundred risks captured, yet easily manage all of the data.

Important: If you have the print version of this guidebook, email the authors for a copy of the Microsoft Excel file: risk@rapidinnovation.com.

Phase 2: Plan the Project

Plan for Success and Failure

At this point you have completed all your planning; your plan is documented and ready for action. Naturally, the first thing to do is implement all the preventive action contained in your plan.

As you recall, mitigation of risk consists of two parts. First take action to ensure that the risk does not happen, and second take action if the risk does happen. Preventive action is all about taking the steps needed to ensure that risks do not happen.

Preventive action should be documented in your plan, and you should not wait for implementation. Your best course is to ensure that the risks do not take place.

LEVEL 1
The Master Process:
Developing and Implementing a Risk Management Plan

Phase 1: Assess

Step 1. Identify Risks

Step 2. Prioritize Risks

Phase 2: Plan and Implement

Step 3. Plan Preventive and Corrective Action (Mitigate)

Step 4. Implement Preventive Action (You Are Here)

Phase 3: Monitor and Control

Step 5. Monitor, Monitor More, and Control

Step 6. Take Remedial or Corrective Action

Phase 2: Plan the Project

Proactive Risk Management

Once you have developed your Risk Management Plan, you move to Phase 3: Monitor and Control. In this phase, you closely monitor what is happening with your project and make adjustments to your plan based on the changes and progress.

This is where the concept of proactive risk management shines. Proactive risk management means that your plan is not stored in a filing cabinet or a hard drive, never to be seen again (which substantially reduces the benefits of the planning effort).

Instead, you schedule regular meetings to monitor and control risks; for example, once per week, once per month or at each project milestone. The meetings should be frequent enough to avoid missing a trigger point. The frequency of these meetings varies depending of the type of project you are managing — use your judgment. The important principle is to have continuous monitoring and control of risks.

Generally these are 60-minute meetings, scheduled solely for the purpose of risk management. We highly recommend that meetings be set to review and manage risks independently of other meetings. We have found that appending risk management to other meetings such as status meetings or project review meetings tends to shortchange the topic. However, you can schedule your risk management meetings right after or just prior to other meetings to take advantage of attendance.

Sometimes it's not practical to have a special meeting; some in your team may feel that there are too many meetings already. Then make your risk review part of a different meeting — for example, the regular project status meeting, or even a design review meeting. Make certain that you ensure sufficient time for risk management; you don't want to have the last five minutes of another meeting, this is not productive. Instead agree that you will have at least 30 minutes; better yet, take a break and let your team members rest. Then start again to talk only about risk management.

One last point: You need to invite key players to your monitor and control meetings — leaders and managers most intimate with the project.

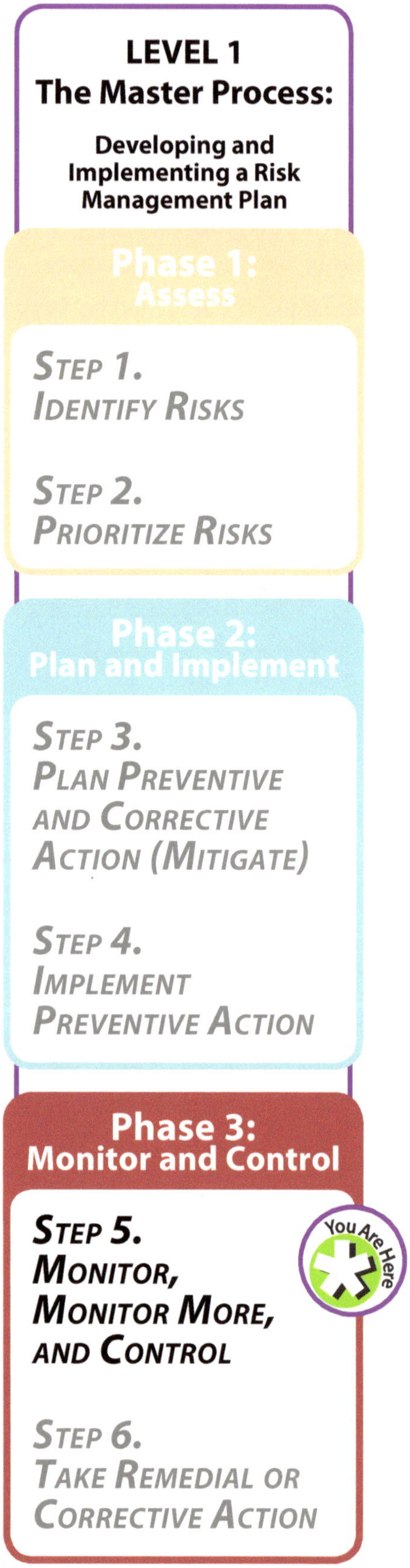

Monitor Your Risk Plan Using the Critical Questions of Proactivity

Template

Agenda for the Risk Management Meetings

The list shown below will provide a starting point so you can develop an agenda that is better suited to your projects. You can add, change or delete items to improve the productivity of your meeting.

We suggest that you project the current version of the Risk Management Plan so everyone can see the changes being made. If you do not have access to a projector, a paper copy of the plan will suffice.

1. What has changed since we last met?
   - ❑ What important changes have taken place that impact our project?

2. How have risks changed?
   - ❑ Are there any new risks that must be added to the Risk Management Plan?
     - What has caused the new risks?
     - What are the implications of the changes?
     - Who are the risk owners for the newly identified risks?
   - ❑ Are there any old risks that are obsolete and must be deleted from the Risk Management Plan
     - What cause the deletion of risks?
     - What are the implications of the changes?
   - ❑ What changes have happened to the current list of risks?
     - What caused the changes to existing risks?
     - Change of Risk Owners?
     - Update to the P and the I?
     - Any new data that add clarity to the risk?
   - ❑ What other changes have take place?

3. What are the impacts to the overall Risk Management Plan?
   - ❑ Are there any trends that will impact our Risk Management Plan?
   - ❑ What will happen if the new trends continue? What should we do right away to mitigate the trends?

4. Validate the needed changes and then document them into the plan prior to the close of the meeting.
   - ❑ What are the specific changes to our current Risk Management Plan?
     - Add new risks?
     - Delete risks?
     - Changes to current risks?
     - Changes in responsibilities?

Phase 2: Plan the Project

Taking the Next Steps

You and your team have been constantly monitoring the Risk Management Plan to identify any trigger points, define new risks and take preventive action to keep risks from happening.

At this point, you have documented what needs to be done in case a trigger occurs. Further, you know who is responsible and you have allocated sufficient resources to handle matters. In short, you are well prepared due to the risk planning you did earlier and your constant monitoring of risks.

When a risk does happen, you are ready to take remedial or corrective action. The purpose of such action is to ensure that your project delivers the anticipated results.

Address the Risks

As you monitor your Risk Management Plan, you will determine that a risk has reached the trigger point — the point documented in the plan at which mitigation will initiate. Take appropriate action, and ensure that those responsible for various activities know it. Your plan already has the specific action you need to take, but you may alter the plan depending on the situation.

Address Risks Not in the Plan

Regardless of the diligence in developing your plan, there is always the possibility that an unexpected risk happens. Because it is not contemplated in the plan, you will need to act fast. Focus on the risk, define a level-headed course of action and lead the implementation. This is where a Risk Management Team comes into play; your team will collaborate with you in addressing the risk and defining appropriate action.

LEVEL 1
The Master Process:
Developing and Implementing a Risk Management Plan

Phase 1: Assess

STEP 1. IDENTIFY RISKS

STEP 2. PRIORITIZE RISKS

Phase 2: Plan and Implement

STEP 3. PLAN PREVENTIVE AND CORRECTIVE ACTION (MITIGATE)

STEP 4. IMPLEMENT PREVENTIVE ACTION

Phase 3: Monitor and Control

STEP 5. MONITOR, MONITOR MORE, AND CONTROL

You Are Here

STEP 6. TAKE REMEDIAL OR CORRECTIVE ACTION

For comments, questions or additional copies please e-mail the author: flexpm@rapidinnovation.com.

The Political Landscape

Introduction

Used in the colloquial connotation, politics is all about human interaction. It is the way we relate to each other and how we persuade each other. While the term "politics" refers to the art and science of government or governing, politics in product development has to do with the way we interact. In this setting, "politics" refers to more at social matters.

Politics can be a negative — for example, to engage in intrigue to get a promotion or to take advantage of peers and colleagues to increase one's power. It can also be a positive; we can think of politics as the lubricant we use to get things done — i.e., grease the wheels. Thus, politics in product development can help or hinder.

We believe that some level of politics is inevitable as long as you have human interaction. It is important therefore for all of us to develop the skills to use politics in the positive fashion to achieve our objectives, while clearly avoid using it as a negative.

We can further define politics in product development as the way to achieve goals through accommodation and the exercise of influence.

Hints to Deal with and Manage Politics

Improve your ability to observe. The power of observation will show you the dynamics of your organization; i.e., how things really get done. Observe how decisions are made, how people behave during meetings, how people relate to each other and how people relate to you. From your observations you can draw conclusions and apply it to your personal strategy for success.

Improve your ability to persuade. Persuasion is all about appealing to peoples' passions and preferences. Persuasion starts with building rapport; that is, finding common ground, or being able to see things from the perspective of the other person. Once rapport is built, try to identify the personal needs of the other individual — again, by looking at things from the other person's perspective rather than yours.

Then, set about creating a win-win situation, where there is gain for you and for the other individual. End by confirming your agreement; this is a critical step in order to cement the decision so you will not need to negotiate the same thing again.

MBWA (Management By Walking Around) is a management style that encourages managers and executives to walk around informally, talking to employees directly, without the need for hierarchy. In chaotic environments, it is important to walk around and talk to your team. Doing this frequently will allow you to understand the politics so you can take appropriate action.

Phase 3: Implement the Project

Execute the Project

In its simplest form, execution is all about implementing your plan, making it happen. In product development the change from planning and executing is not as clean. In most cases, you will be finalizing your planning while at the same time you will be implementing; again this is due to the nature of product development and pressure to deliver the new product.

Your plan will also change as you proceed with implementation, During this phase your ability to use "proactive project management" will help you achieve the results.

At this point you will have a WBS completed or near completion. You will also have defined a schedule, roles and responsibilities and a budget. Also, you will have completed the Kick-Off meeting, which should have added much clarity to your ability to execute.

Phase 3: Implement the Project

Rolling Wave Planning

Because of the confusing nature of product development, Rolling Wave Planning is a very practical way to implement your plan. The methodology requires you to have absolute clarity of the steps towards implementation for a short time frame in the immediate future. For example, you must have a complete description of all the activities needed for the next three months (which is what we recommend). The actual time frame depends on your needs. You could choose a window of two months, or perhaps four. The important step is to select a brief period that covers the near future. The rest of your plan can have uncertainty.

This window keeps moving. As one month goes by, your plan covers the new time-frame. For example, in January a three month window includes January, February and March, but on February first the window now shows February, March and April. Thus, the idea of a rolling wave.

This form of planning allows for the uncertainty of product development, but keeps the team working on the right things all the time. It ensures maximum progress without the need for a complete plan. In product development this becomes very practical due to the complexity of developing new products.

Important Notes

- You must have considerable detail for the selected time frame — e.g., every task must be clear, dates must be accurate.
- You must constantly review your Rolling Wave Plan. This is done during your project review meetings.
- There is always a moving window of time; this window is selected by you and the members of your team. We suggest three months, but you may use a different length.

Example: Rolling Wave Planning

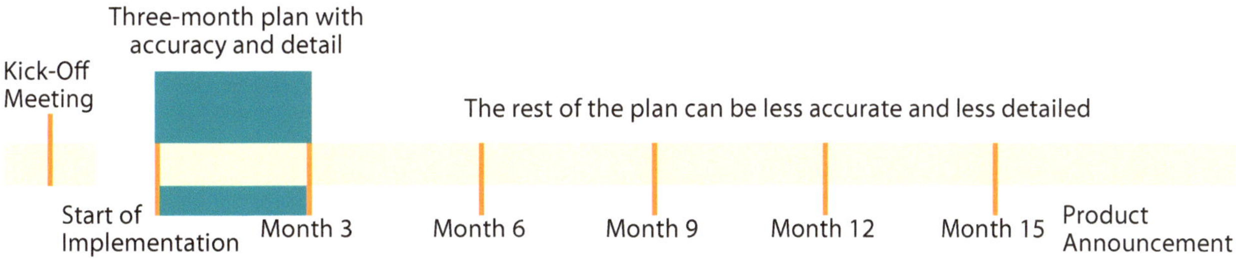

This diagram captures the strategy behind Rolling Wave Planning. In this example, there is a three-month window where much detail and accuracy is included (while the rest of the plan remains less clear).

Phase 3: Implement the Project

Review (Status) Meetings

> Essential
> "Review meetings are all about the future."

The implementation phase of any project relies heavily upon regular review or status meetings. This is the one area where every aspect of the project is evaluated to ensure that the entire project is on track to meet its objectives.

You need to work with your team to decide how often these review meetings take place. In some cases it may be every day; for example, you are working on a very complex project. In other situations every week or every other week is optimal. You can change the frequency of these meetings, but keep in mind that one of the benefits of a meeting regularly is reliable attendance.

You may wish to schedule a major review when the project is evaluated in full detail, perhaps every quarter or twice per year will work for you. You can also call an emergency review meeting in order to deal with an unanticipated risk, a breakdown or a major change in your project.

Review Meetings and Multi-Site Development

There is a challenge of communication when several languages, cultures, sites and time zones are at play. In such circumstances, review meetings take on special importance. Many companies schedule video-conferences for their regular reviews, and a face-to-face meeting for the major reviews.

We highly encourage you to travel to all the sites involved in your project with regularity — a small investment compared to the benefits. It's painful, if not impossible, to manage a project without face-to-face interaction.

The Agenda for Review Meetings

Using Proactive Project Management, review meetings are not about what has happened since the last meeting nor what is happening now. Rather, they are all about the future. Use these meetings to learn what's happening insomuch as it enables you to make informed decisions about what you will improve or continue in the future.

1. What is the variance between the initial plan and the current situation — i.e., today? Cover this for schedule, budget, objectives, etc.
2. What are the causes of the variance from the initial plan — i.e., the root cause of the changes from the initial plan? You can also ask the question using your last review meeting as the reference.
3. What is the current trend if we do nothing?
4. What specific actions should be taken as a result of the variance to bring the project back on track to completion? The focus is the ultimate goal of the project — in your case, to deliver a new product to the market by a certain time, at a certain price and with a certain profitability.
5. What new risks have emerged since our last meeting?
6. What action should we take to either avoid or deal with risks if they do happen?

Important Points About Review Meetings

- Avoid the risk of your review meetings becoming too routine. Your team may have the tendency to see the review meetings as required but unimportant. To avoid this, use the agenda outlined above. And remember that these meetings are all about the future.
- Document every action item very carefully during the meeting. Ensure that all attendees knows their responsibility once they leave the meeting. For every action item, document the date of expected completion and the name of the person responsible for the action.
- Follow up with your team members after the meeting to ensure that they have the necessary information to carry out their action items. Remind them to report progress or completion at the next review meeting.

For comments, questions or additional copies please e-mail the author: flexpm@rapidinnovation.com.

- Always connect your review meetings with your status reports; that way there is a single message. Synchronizing both will avoid confusion.

Status Reports

Important: Just because you emailed a status report doesn't mean that everyone is informed!

Keeping everyone informed in a chaotic environment is very difficult. You will have to work hard and try many ways to do it. Consider that people in product development receive a considerable amount of information from many sources. Also, remember those in your team who are not collocated with you, perhaps separated by time zones, language and culture.

The overall intent is to keep all stakeholders informed so they can do their jobs. The status report becomes one of tools you will need to keep every informed about your project. Remember, just because you emailed a status report doesn't mean that everyone is informed!

Definition

A status report is a document, typically one page long, that highlights and briefly describes the project's status. Generally, it contains the variance of scope, cost, time and quality. It also shows any major accomplishments such as the achievement of a critical milestone. In addition, it summarizes issues, predicts trends and shows actions required to overcome issues and reverse any negative trends.

Status reports are not about the history of the project — rather, they are all about the future of the project.

Why Status Reports

Ensuring clear communication — i.e., the status of your project — is difficult. Do not underestimate the challenge! This guidebook includes an extensive section on advanced communication, and we strongly suggest you review it and apply the principles to communicating the status of your project.

Status reports enable you to communicate clearly and unequivocally!

Above all, the status report process is to keep the project's customers and stakeholders fully informed. This reduces stress and encourages engagement by everyone.

Key Questions

- What is the variance between the baseline and the current status of the project?
- What key milestones were achieved since the previous status reports?
- What major accomplishments need to be recognized?
- What is the status of previous issues? What issues have been resolved?
- What is the variance between the planned budget versus actual?
- What are the issues that are causing the variances?
- What risks may pop up in the future? How are we going to mitigate future identified risks?
- How did we deal with prior identified risks?
- What is the projected trend for scope, budget, quality, results and objectives?
- What action is required, and by whom?
- What are the requests for management?

Phase 3: Implement the Project

Status Reports and Multi-Site Teams

It takes considerable more energy to keep your team, managers, supervisors and other stakeholders informed when the team is located in more than one site. Especially if more than one language and culture are involved.

We suggest that you tailor your status report to the needs of your most distant stakeholders; this will help you realize the extra effort needed to keep your entire team informed.

Good Practices for Multi-Site Status Reports

- Hold quarterly or half-yearly project review meetings preferably at the farthest site involved in the project.
- As a project manager, travel to all the locations at least once per year to personally understand the current status of your project.
- Ask people from your company who are traveling to the other sites to hold project review meetings on your behalf.
- Make wise use of technology. While video-conferencing is very helpful, it is no replacement for face-to-face contact.

Technology and Status Reports

Use Technology Wisely

Studies have shown that words alone represent only about seven percent of communication; the rest is communicated through the voice, gestures, facial expressions, etc. Given this premise, email is only words; consequently, it is not the best form of communication, particularly when it comes to complex subjects such as decision making.

The following table shows the more common means of communication available to project managers and what is lost in communication by each method. While it is not possible to meet frequently in a multi-site environment, we suggest that more face-to-face meetings are possible. Only face to face meetings provide all the stimulus for good communication. We suggest that you use technology wisely to apply the right tool for the right communication task.

Type of Communication	Words	Voice Inflection	Immediate Feedback	Body Language	Touch and Proximity
Face to face meetings	YES	YES	YES	YES	YES
Video conference	YES	YES	YES	YES	
Peer-to-peer video	YES	YES	YES	YES	
Telephone	YES	YES	YES		
Chat rooms	YES				
Email	YES				
Fax	YES				
Intranet, blogs and alike	YES				

Example: Status Report (Progress Report)

Project Name: ______________________________ Date: ______________

Project Leader: ____

Item	Mar	Apr	May	Jun	Jul
Visit customer site.					
Prepare the plan.					
Develop costing.					
Obtain initial permits.					

Current Issues: ______________________ Response: ______________________

Risk Management: ______________________ Response: ______________________

Action Required: ______________________ Response: ______________________

There are many formats for status reports. Your company may already have one. If so, you need to use it. The format is not the important factor; the ability to communicate clearly and unequivocally the status of your project is.

In the example above there is a small Gantt chart representing the current time window. This helps your reader understand the time frame. There is also a list of current issues. If you don't already have an Issue Tracking Process, then you may omit the lsit of current issues from your status report or keep it in your status report to communicate new or large issues. There is a section for risk management. You can list emerging risks or risks that have taken place and have become issues.

Always make your status report action-oriented. Everyone needs to know what he or she needs to do. Your status report is all about the future or your project, not the past. This is because regardless of the past you still need to deliver results.

Phase 3: Implement the Project

Communicating Status

Target Your Reports to the Audience

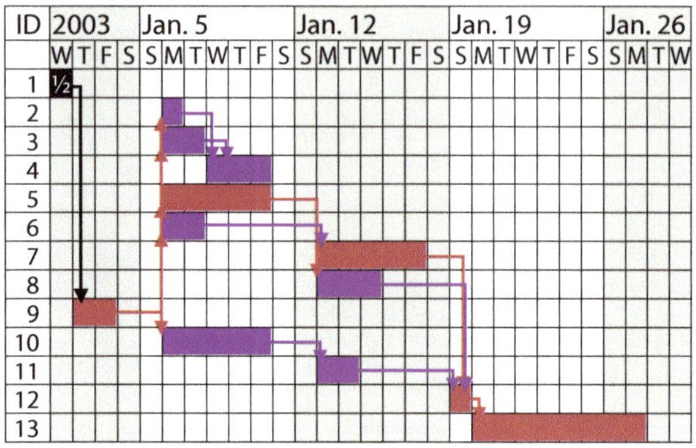

Bar Chart

Milestones for Project Fuzzy Wave

ID	WBS	Milestone	Date
1	1.1.0	Project Authorized	Thu 1/2/03
12	1.3.4	Approval to Proceed to Design Phase	Fri 1/17/03
22	2.3.1	Power-On Prototype	Fri 3/14/03
34	4.1.2	Start Pilot Run	Tue 4/29/03
46	5.1.0	General Availability (GA)	Thu 7/10/03

Milestone Chart

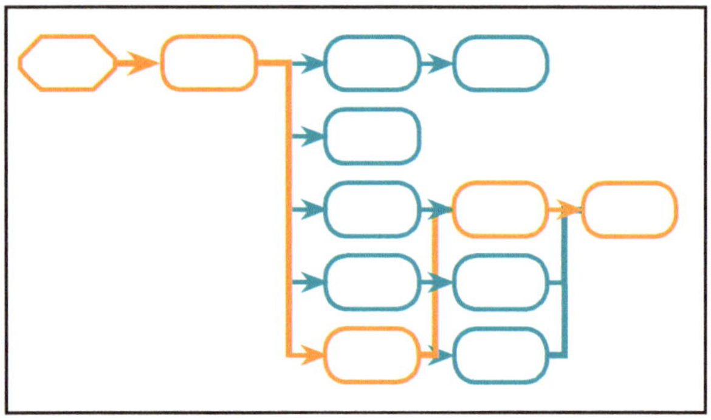

Network Diagram

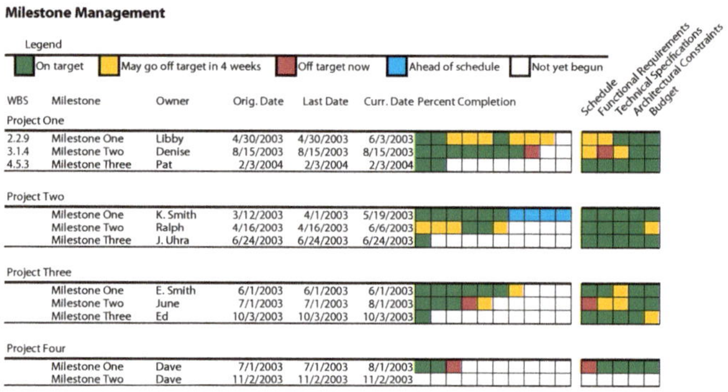

Dashboard

Type	Simple	Dependencies	Visual Time	Level of Detail	Use
Bar	😐 NEUTRAL	😐 NEUTRAL	🙂 GOOD	MEDIUM	Determine team status.
Milestone	🙂 GOOD	☹ BAD	☹ BAD	LOW	Management reports.
Network	☹ BAD	🙂 GOOD	☹ BAD	HIGH	Detailed team planning.
Dashboard	🙂 GOOD	☹ BAD	☹ BAD	LOW	Executive updates.

Table

The many diagrams on this page should remind you that your audience members are different — some are engineers, others are executives or managers and yet others are in marketing and finance. You need to customize your reports to your audience so that each person can understand it. You need to understand the "language" spoken by your audience. For example, executives and managers speak the "language of dollars," marketing people speak the "language of benefits" and engineers speak the "language of precision."

Scheduling: The Jogging Line

The jogging line spells out the amount of time each project activity is ahead or behind the baseline schedule. The jogging line therefore indicates the fraction of completion for each activity and what remains to be completed.

In recent application, the jogging line has been viewed as a step in proactive management of schedules — in particular, the amount of time each activity is ahead or behind the baseline schedule. This can be used to predict the project completion date and map any corrective action needed to align all activities to meet the deadline.

Notice that the jogging line is a visual tool. It will serve you best if you can project it on the wall for all to see. As an alternative, you might want to use a wall-size version of your schedule, and hand-draw the line.

In addition, the jogging line is a point-in-time tool. It shows where all activities are at a particular time. Think of it as a snapshot view of your project.

Important Notes About the Jogging Line

- A new Jogging Line is created every time you have a project review meeting.
- The shape of the Jogging Line will change with every project review meeting to reflect the changes since the previous meeting.
- The Jogging Line is best used with the Milestone schedule — the high-level Gantt chart.
- It is a team tool — not an individual tool! The Jogging Line is a visual tool, as such it must be displayed for the entire team to review and study.
- After the Jogging Line has been reviewed, you must create an Action Plan with clear responsibility to ensure that corrective action is taken to bring all late items up to date.

Phase 3: Implement the Project

Example: Hierarchical Scheduling

Master Schedule — Drives Every Other Schedule

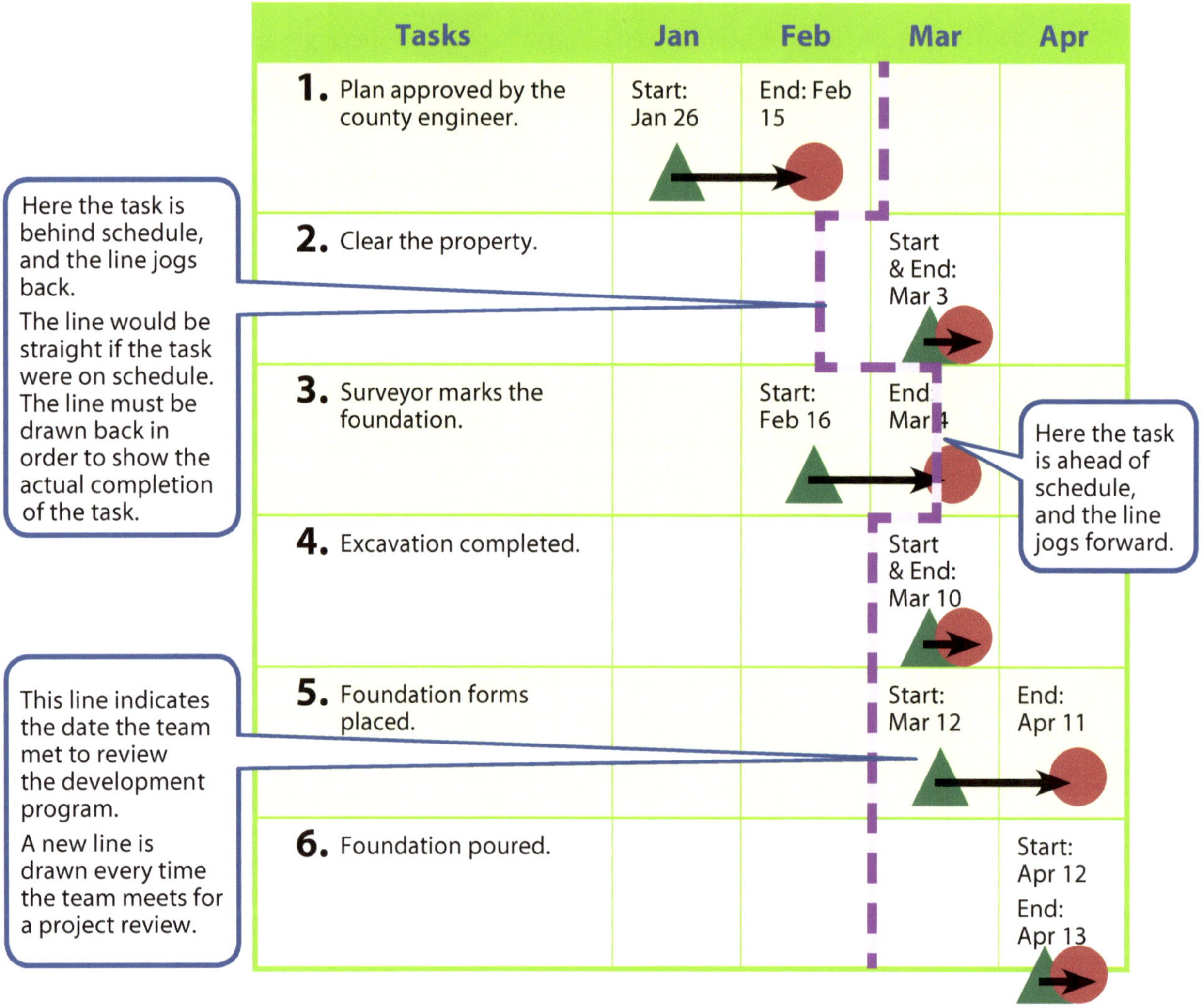

Phase 3: Implement the Project

How to Develop the Jogging Line

The jogging line approach to managing your schedule changes with time; that is, the vertical jog must be drawn every time you have a review meeting. As time moves on, every new jogging line will appear forward of the previous one in order to match the date of your review meeting.

The jogging line is a powerful way to visually show the status of your project. It works best when done in full collaboration with your team. It is also a good tool to communicate your schedule to management by using the major milestones of your project; that is, not the detailed plan.

We suggest that you post the latest jogging line on a bulletin board or make it available as a web page for every member of the team to understand that status of the development project.

Detailed Process: Develop the Jogging Line

1 Set the Baseline Schedule

Start with the project's baseline schedule — i.e., the schedule you are using as the final or reference.

2 Schedule the Project Review

Mark on the schedule the date when it will be reviewed — for example, the day of the project review meeting.

3 Interview Contributors to the Schedule

Proactively interview all persons who contribute to the schedule — that is, the owners of each activity. Determine if their activities are ahead, on time or behind schedule, and by how much. This must be an honest and guilt-free conversation, as you want to get information that's as accurate as possible.

4 Draw the Jogging Line

Then, draw a vertical line from top to bottom. Jog the line to show at that particular point how much of each activity has been completed (the actual completion).

You will have only three choices:

- The activity is on time, and the line is vertical.
- The activity is behind schedule, and the line will jog to the left.
- The activity is ahead of schedule, and the line will jog to the right.

5 Predict Completion

Complete a jogging line for each element (activity) in your schedule. Now you can predict project completion and any corrective action needed to bring those activities that are behind schedule up to meet your completion date.

Learn and Improve (Kaizen)

Review your jogging line during the meeting that you previously scheduled

Take appropriate corrective action.

Phase 3: Implement the Project

Scheduling: The Slip Chart (Wiggle Chart)

Introduction

The Slip chart — also called the Wiggle chart — is part and parcel of proactive project management. It focuses on schedule control. It anticipates the expected rate of future project progress. Unlike other proactive schedule control tools, it focuses on the predictions of major project events, milestones and project completion.

In the examples below, the vertical axis shows the team's predicted completion date of a specific milestone, while the horizontal axis shows the actual date the prediction was made — i.e., the date when the prediction was updated. Thus, the beginning point on the horizontal axis is the time when the schedule baseline is prepared at the start of the project, and the milestone dates are marked on the vertical axis. After the beginning point the project work is kicked off, and the horizontal axis represents the actual project timeline; the team reviews progress regularly and makes milestone predictions. By connecting all the predictions for a particular milestone into a line, we can obtain the milestone trend line. If the line approaches the completion line going upward, we know that there has been a milestone slip.

Delivering the milestone right on time would produce a line approaching the completion line horizontally (Example 1). If we estimate an early milestone completion, the completion line would trend downward. Although it is effective in predicting milestone progress, the chart is even more effective if used to develop the actions required to eliminate any potential deviation from the baseline milestone schedule.

Examples: Scheduling with the Slip Chart

Example 1

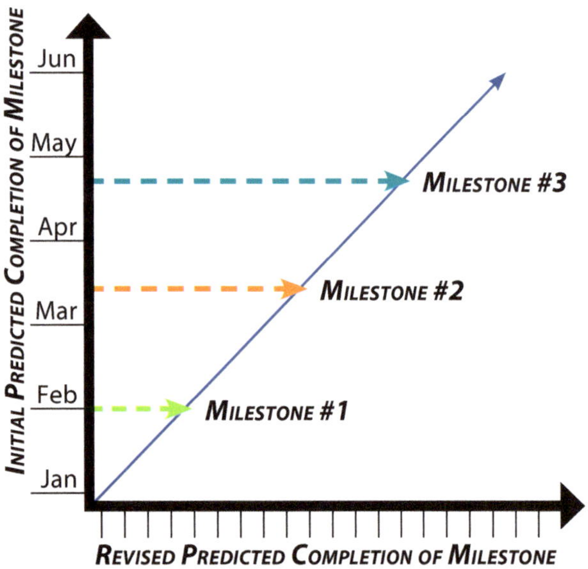

Example 1 represents a hypothetical example if everything went well. All the horizontal lines would be straight and not a single milestone would be missed.

Example 2

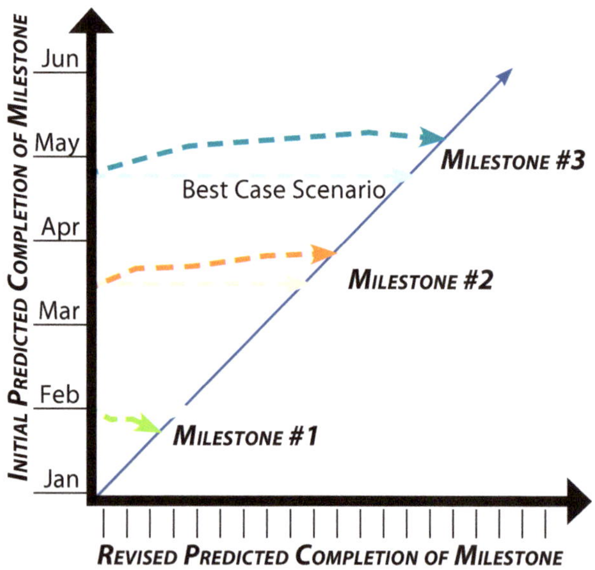

Example 2 is more realistic than Example 1. Notice that Milestone #1 was early, but Milestone #2 was late, which caused Milestone #3 — the overall project — to be late.

For comments, questions or additional copies please e-mail the author: flexpm@rapidinnovation.com.

Phase 3: Implement the Project

Example 3

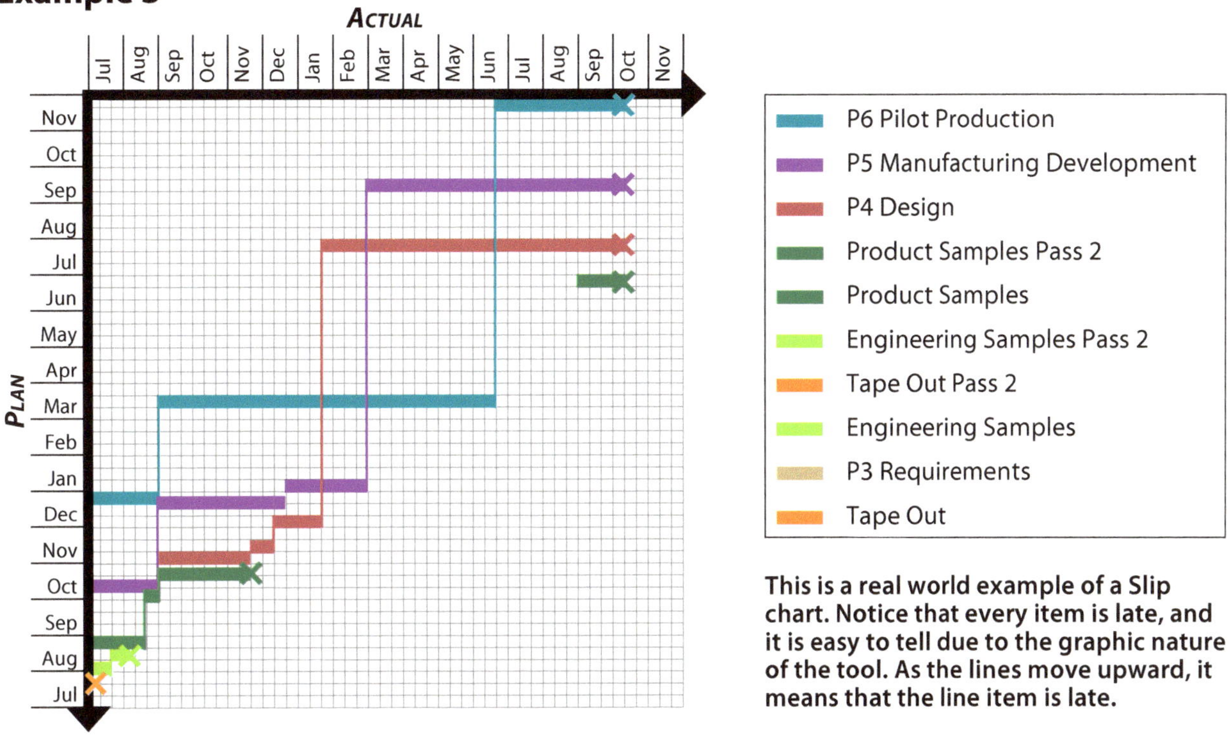

This is a real world example of a Slip chart. Notice that every item is late, and it is easy to tell due to the graphic nature of the tool. As the lines move upward, it means that the line item is late.

Example 4

ABC PROGRAM SLIP CHART

DAYS	RISK DAYS	PILOT BUILD
Remaining Days:	XX	YY
Days Planned at DP:	XX	YY

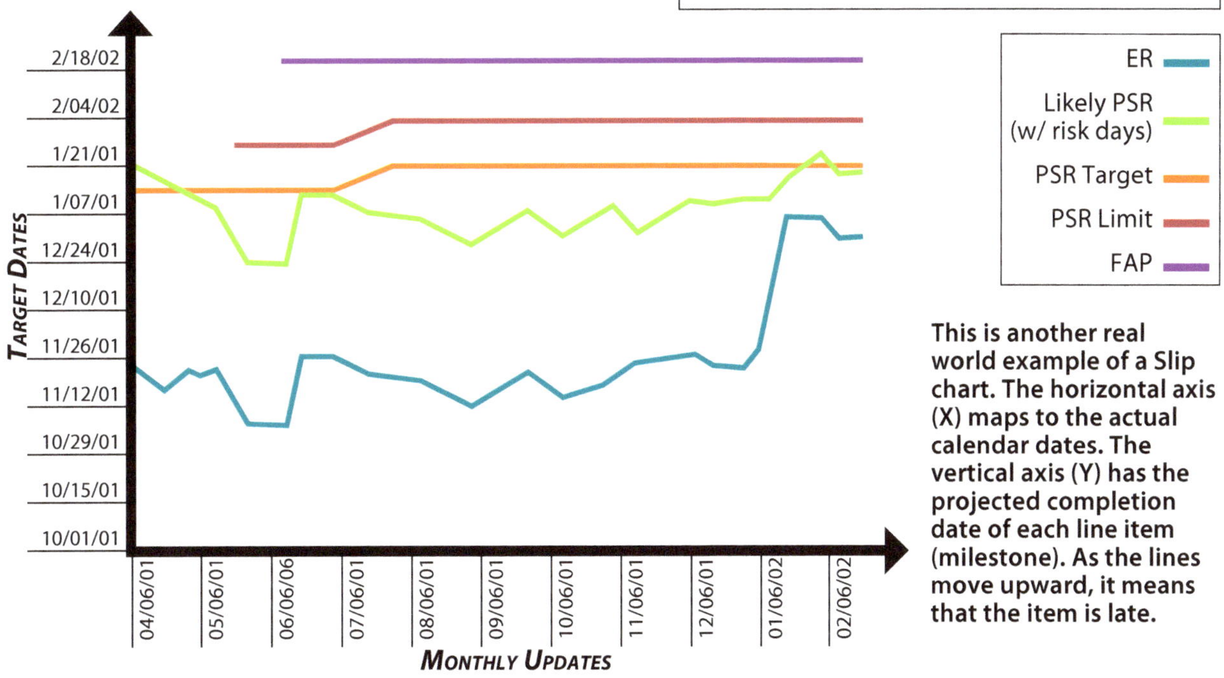

This is another real world example of a Slip chart. The horizontal axis (X) maps to the actual calendar dates. The vertical axis (Y) has the projected completion date of each line item (milestone). As the lines move upward, it means that the item is late.

Phase 3: Implement the Project

Example 5

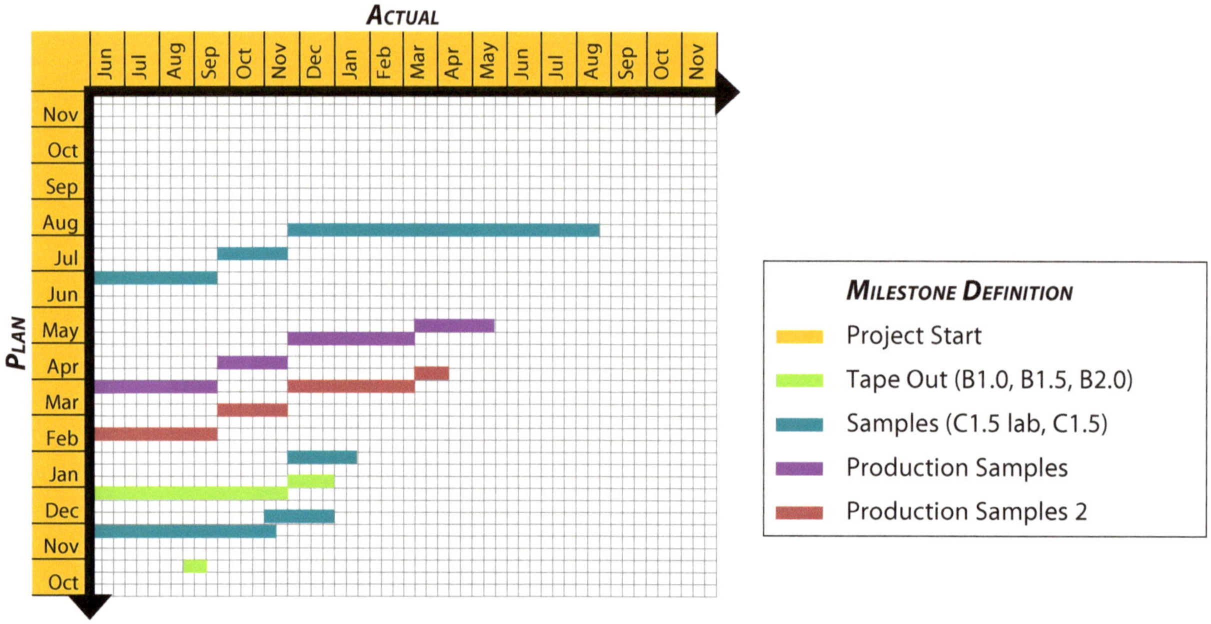

In the real world example above, we see a slip chart from a project in the semiconductor industry. At the top of the chart is the entire project (the November row) or the last task to complete prior to completion of the project. Below are all the major milestones, starting with the earliest one at the bottom of the chart.

When the lines are updated and they move upward it means that the task is late. When the lines move downwards it means that the task is early (ahead of schedule).

By plotting all your milestones from the earliest to the latest, and updating them often, you can tell if your project will be on time or late.

Phase 3: Implement the Project

Detailed Process: Slip Charts

1

Collect Needed Information

Identify the sources of information that will help you in developing the Slip chart.

For example:

- Project scope.
- Schedule — i.e., Gantt chart or any form of timeline.
- Key assumptions for your project.
- Key success factors.
- Constraints and risk analysis

Important

If slip charts are not agreed to by your organization, then you must first ensure consensus, then prepare a deployment plan for the adoption of this tool.

2

Create the Chart

Key steps:

- Start by plotting all the key milestones on the vertical (X) axis. Note that it is best to use milestones as it keeps the chart more universal, allowing you to use it with managers and executives.
- The first (top) milestone must be the completion of the entire development project.
- Work your way down by plotting each subsequent milestone until you get to the start of your program; i.e., the earliest milestone.
- Label each milestone with its name and the projected completion date.

Important

You slip chart should have few milestones, generally no more than 10, and preferably fewer. Too many items makes it difficult to understand and distracts from the strategic intent — i.e., delivering results.

Important

The first time you plot your chart, you may realize that information is incomplete. Fear not; lack of information is natural at this point, plus it tells you what additional information you must obtain.

3

Meet with Stakeholders to Validate Your Chart

This is the first meeting.

These meetings are not trivial or short. Careful dialog and analysis are needed to ensure accuracy and accountability.

Call a meeting with the owners of all key milestones, plus other stakeholders. These are the people who will make or break your schedule.

The purpose of this meeting is to validate all milestones, inform all stakeholders and continue to create ownership of the schedule.

Review and validate all milestones. Make corrections as needed, and document any changes to the assumptions, risk, etc.

Continued

Phase 3: Implement the Project

Cont'd

Important

Ask the Five Questions of Proactive Project Management

1. What is the variance between the baseline and the actual schedule?
2. What are the issues causing the variance?
3. What is the current trend if we continue with the current performance?
4. What new risks may pop up in the future, and how could they change the predicted completion date?
5. What specific action should we take to prevent the predicted completion from happening?

Another hint: Forward the questions to the attendees prior to the meeting.

4

Progress Review Meetings

Once your baseline has been established and your project is off and running, you need to periodically call Progress Review meetings.

This step in the process marks the first of these periodic meetings.

As in step 3, these meetings are neither trivial nor short. Careful dialog and analysis are needed to ensure accuracy and accountability. Invite the milestone owners and other key stakeholders — anyone who can make or break your schedule.

These are special meetings where the focus is schedule control. While you may have project review meetings, status meetings, etc., these progress review meetings deserve focus, and the right individuals.

5

Conclude Your Meeting with an Action Plan

Plot the new dates — i.e., forecasted completion of all milestones. This is where the Slip chart starts showing its value. As you plot you will see items moving higher (late) and items moving lower (early).

In a perfect world, all items should be perfectly horizontal.

Clearly define the action required to bring your schedule under control. Include roles and responsibilities, plus dates when action is expected.

Schedule the next progress review meeting.

Document and archive relevant information.

Follow up on the action items to ensure completion prior to the next review meeting.

6

Track Progress

Key steps:

- Publish the updated slip charts, action items, etc.
- Report to management as appropriate.
- Track progress by meeting with individual milestone owner.
- Call the next progress review meeting.

7

Call the Next Review Meeting

Repeat the entire process every time you have a review meeting.

Learn and Improve (Kaizen)

As you use the Slip chart, you will find ways to improve its accuracy, clarity and usefulness.

Change Order Management

The impact of a project's changes on its scope, schedule, cost, quality and other matters may easily surpass the awareness or expertise of the project's originator. As a consequence, the project may suffer severely or, in certain cases, even collapse. For this reason, it is very important to ensure that each change is evaluated in a disciplined and professional manner before receiving a go-ahead. It is the task of the project change request to help perform an all-angle evaluation of proposed changes.

Critical Distinctions

What is change in the context of Project Management? Change is a departure from the agreed scope of the project. In some cases, it is a departure from the agreed design specifications.

Change is everywhere and happens constantly. Given this complexity, it's important that you establish the various change mechanisms for your development project.

Some authors recommend a single process to handle all changes in product development, from engineering changes to marketing and manufacturing. We believe this approach is suboptimal because so many changes are small and stay within a small group of people.

Electronic tools for change management are becoming more useful and capable, but even these face the complexity of the type, frequency, origin and impact of the change.

Sources of Changes

With change order management comes Engineering Change Notices (ECNs) or Engineering Change Orders (ECOs). Generally, these are internal to the engineering organization, but often they impact documentation, manufacturing, QA and other functions. Be certain to develop a process that reflects the other organizations which might be impacted by these change notices.

In software development there are bug tracking changes, code changes, etc. This is a challenge for the software organization as there are many electronic tools to manage changes to the software. The escalation protocol is critical — many changes are minimal and don't require much review, while others may impact the very product.

Changes to the product definition generally come from the marketing organization; although, a technology obstacle may cause a change in the definition of the product. The MRD (Marketing Requirements Document) is called for. There must be a clear, well documented process to any changes to the definition of the new product. This is one area of weakness is many NPDs (New Product Development Processes) — changes happen ad-hoc, which have later negative impact to your schedule and the new product itself. It is very easy for a marketing person to go directly to one engineer and make a change. Instead, the strict protocol must be established, with a corresponding process map to ensure integrity. This is more so, when you are leading a multi-site development project.

Finally, there are changes to your project that should also be made using a protocol and a documented process.

With so many ways to account for change, how can you keep track of it all. First, this is where your regular project review meetings help a great deal. Add an item to the agenda to review any changes, from any part of your project.

Create escalation protocols that clearly outline where the decisions for a change are made, and who needs to review and approve. This is ideal, but the criteria for escalation may prove daunting. Work with your team to address them. In the end though, judgment will play a critical role, and the sub-teams and individuals in your organization need to be empowered, not afraid to raise an issue or ask questions.

Phase 3: Implement the Project

Change Order Management in Product Development

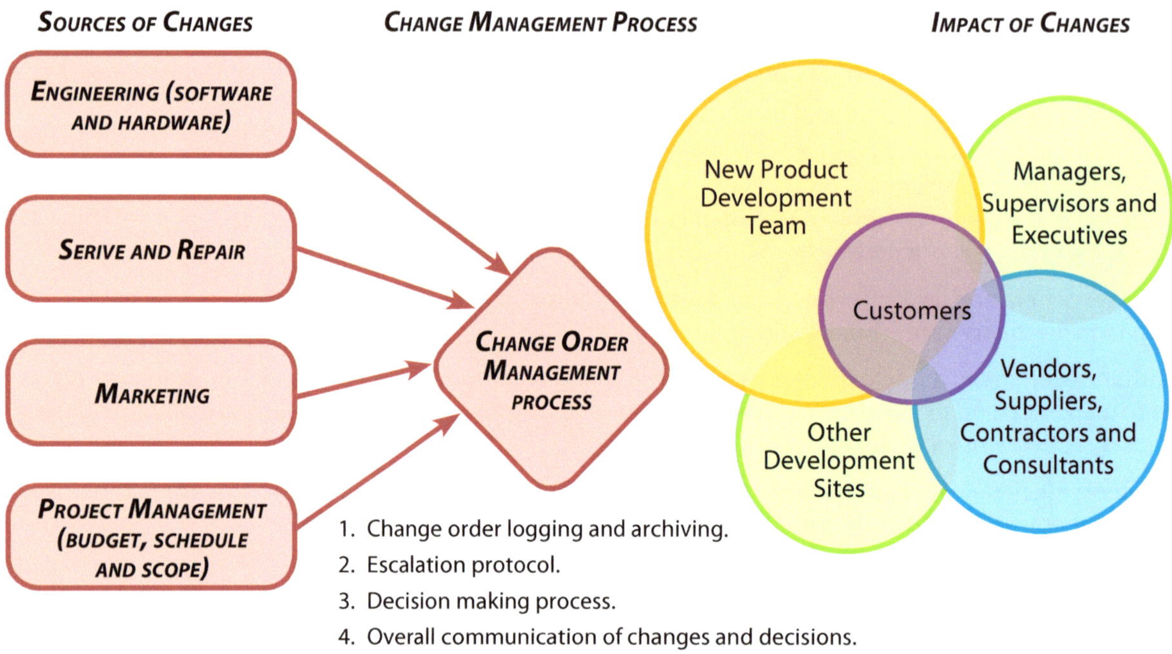

This conceptual diagram illustrates the challenge in managing changes in a product development environment. The sources of changes are many. Then comes the decision to accept, modify or reject a change — shown as the Change Management Process. Finally, the most difficult task, is communicating the decision to all who must know — other sites, vendors and suppliers are frequently left out of Change Order Management Processes.

While there are electronic tools to help you mange changes, in the end you and your team need to proactively collaborate to make the process work.

Beware of Scope Creep

Scope creep, or uncontrolled change of scope, is often perceived as a major threat to any project. While change is ever-present in the business world, uncontrolled change is bad news. Thus, change by itself is not bad, it is part and parcel of today's business world. It is uncontrolled change that creates the problems.

Step One: Establish the Change Protocol

Establish the change request process protocol. (See Protocols in this guidebook.) Your project team, as well as all the key stakeholders, must agree and follow a change protocol. We recommend you develop the protocol in collaboration with the project customers, your project team and key stakeholders. It does little good if you are the only one who understands the protocol — everyone involved must be aware of it and committed to adhering to it.

Example: Project Change Request

Project Name: EL DORADO Date: ____

Project Leader: Frank Morales

Detail each change requested and its impact on the project scope and quality:

1. Change the size of the main control knob on the front panel. The project will be impacted by seven days due to tooling changes. Change is needed for ease of use, customers cannot operate the current knob because it is too small. Changes are contained in new mechanical drawing #ME-1234--900-08
2. ____
3. ____
4. ____

Name of person originating the request: Ann-Marie O'Mally, ME, Shanghai Design Center, ann-marie@company.com

Describe any emergency action needed: Frank Morales

Type of change: ☑ Minor — requires no planning and minimal impact on the budget

☐ Major — requires planning and possible high impact on the budget

Is the change customer funded: ☐ Yes ☑ No

Any additional relevant information: Ann-Marie O'Mallie has already contacted the supplier of the new knob and determined that it can be done in the time suggested. Vendor needs official change notice from our Purchasing Agent

Approvals ____

Phase 3: Implement the Project

Detailed Process: Change Order Management (ECNs)

1

Establish the Protocol for Managing Change Orders

Questions to answer:

- Why do we need change orders for our project?
- What is a change for our project? What constitutes change?
- Who will track all changes?
- Who will keep a log of all changes?
- Who will approve a change?
- What are the approval or rejection criteria? (For example, what will be the impact on the budget, timeline or customer satisfaction?) What is the degree of urgency or type of change?

2

Develop and Start Using the Change Order Process

3

Communicate the Process to Everyone

Checklist:

- Your team members.
- Key stakeholders such as managers and sponsors.
- Project customers.
- Vendors and suppliers as needed.
- Contractors.
- Consultants.

4

Change Order Initiated and Submitted for Approval

A

No

The change was not approved.

B

Not Yet

The change is not yet approved. Either more information is needed or there is a delay.

C

Yes

The change was approved.

5

Close

Checklist:

- Inform everyone.
- Enter in the log.
- Archive.

5

Implement, Track and Complete

Checklist:

- Inform everyone about the impact on the schedule, budget and the rest of the project scope.
- Document impact of the change on the schedule, budget and the rest of the project scope.
- Enter into the log.
- Track the progress of implementation.
- Verify completion.
- Close the change order.
- Archive.

Phase 4: Close Out the Project

NEW TOPIC

Complete and Close Out the Project

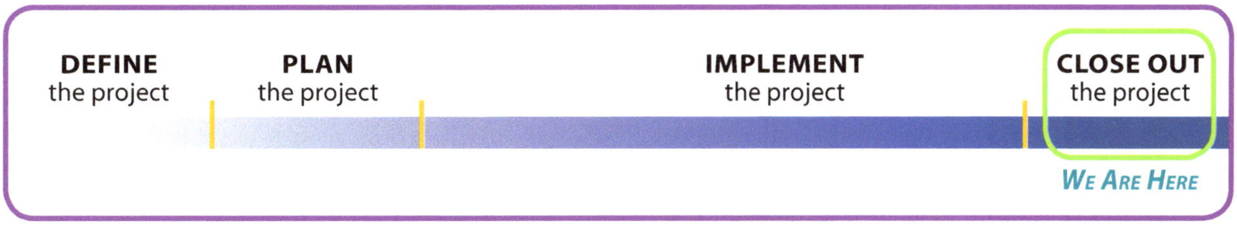

In product development, the end of the project is when the product is announced to the market place (NPI). The actual day varies from company to company. Some mark the day when the product is available to the customer to purchase; other companies allow a few weeks or even a few months before the project is considered closed. Once the project has ended, it officially becomes the responsibility of marketing through Product Management or Product Line Management.

Just as important as defining the exact start date, you must clearly define the date when your project will be considered closed. Team members need to move to other projects, review meetings need to end, other activities need to be terminated, etc. We strongly recommend that you use Debriefs. We've included a comprehensive section on the topic.

Debriefs

Debriefs by Many Other Names

First, we must clarify the name of this methodology. It goes by several names depending on the industry and function you represent. What we will refer to throughout this guidebook as debriefs" is also referred to as postmortems, postmortem reviews, retrospects, retrospectives, lessons learned, after action report and post-project reviews. We recommend that you use whatever name you have used in the past and forget about the others to avoid unnecessary distraction.

Purpose

The end goal is to obtain information from projects in order to improve the future productivity of the organization. A debrief is all about creating the capacity for action to improve future performance. It is about a search for knowledge and opportunities. Its purpose is to create learning for the individuals in the organization to sustain positive improvements and to improve future performance.

See Also
Debriefs and Postmortems for Product Development

Essential

As part of this series, we have developed a guidebook that deals exclusively with the debriefs in product development. Email the authors for additional information.

Phase 4: Close Out the Project

The Business Case

In the words of the famous Spanish philosopher George Santayana: "Those who cannot learn from history are doomed to repeat it." Organizations cannot afford to keep repeating defects and mistakes due to their impact on the bottom line. They also interfere with your ability to compete in an ever more complex world. Neither can organizations overlook innovation nor opportunities for improvement derived from well-orchestrated debriefs.

Product development is a chaotic environment where constant change is the rule. Further, product development is becoming more complex — for example, multi-site product development, strategic partnerships, co-development and much more. All of this is uncharted territory, not the ordinary process of years past. This new environment requires constant examination though a robust framework of debriefs and their corresponding improvement plans.

Finally, over the years, we have validated that debriefs are a fine tool for team building. They improve morale, re-energize the development teams and foster a collaborative environment. This is due to the consolation that problems will be addressed and good practices will be repeated. Debriefs create a vivid environment of continuous improvement and the reassurance to your development team that things will continually improve.

We estimate that a well-orchestrated debrief can improve productivity by one percent. Four debriefs per year is our recommendation. Just imagine, in one year the productivity of your development team could be up by four percent! Now imagine conducting debreifs for five consecutive years — you could improve organizational performance in the range of 20 percent!

Strategic Intent

The intent of every debrief is to answer several critical questions about a specific project. Every debrief must conclude with a plan of action: What must we do differently next time?

Strategic Questions That Debriefs Answer

- What was supposed to happen?
- What actually happened?
- Why were there differences?
- What must we do differently next time

This set of questions explores deviation from the plan. Deviation, while not inherently wrong — particularly in the tumultuous environment of product development — may indicate an opportunity for improvement.

The process isn't complete until you understand "why" it happened.

Tactical Questions That Debriefs Answer

- What worked?
- What didn't?
- Why?
- What must we do differently next time?

This version is more tactical. The focus is still on deviation from the plan, unanticipated results and the root cause of the change. But these questions also suggest what changes to future behaviors need to be made to produce a desirable outcome next time.

Key Assumptions

For debriefs to work on a continuous basis, the following assumptions must be true:

- The information that was collected and prioritized is put to use to improve specific processes, procedures or tasks.
- Debriefs are an on-going, enduring undertaking for your organization, not an occasional, isolated activity.

Important

Debriefs have been shown to be the best way to improve the performance of an organization.

Phase 4: Close Out the Project

- There is a documented process to deploy the information and to incorporate it into the practices of your organization, which will result in measurable performance improvement.
- There is a small and simple set of metrics to track progress.
- Debriefs are in actuality and perception non-punitive and guilt-free.

Benefits

Studies have shown that the best way to improve the productivity of any organization is through a robust program of debriefs. They allow the team to capture real information from actual experience. This information is critical to improve the performance of future projects.

Debriefs are well-orchestrated activities that capture targeted information from a development team for later use to improve the organization's overall performance. This information can be lessons learned, key learnings, process improvement and any other data or insight that can be used to reach higher levels of excellence.

Debriefs should be part of the continuous improvement strategy of the organizations (Agile, Kaizen, ISO 9000, Six Sigma, etc.). Debriefs foster the habit of learning from experience, which helps create a learning organization. The diagram below illustrates some of the benefits of debriefs. The list is not complete as the benefits are many.

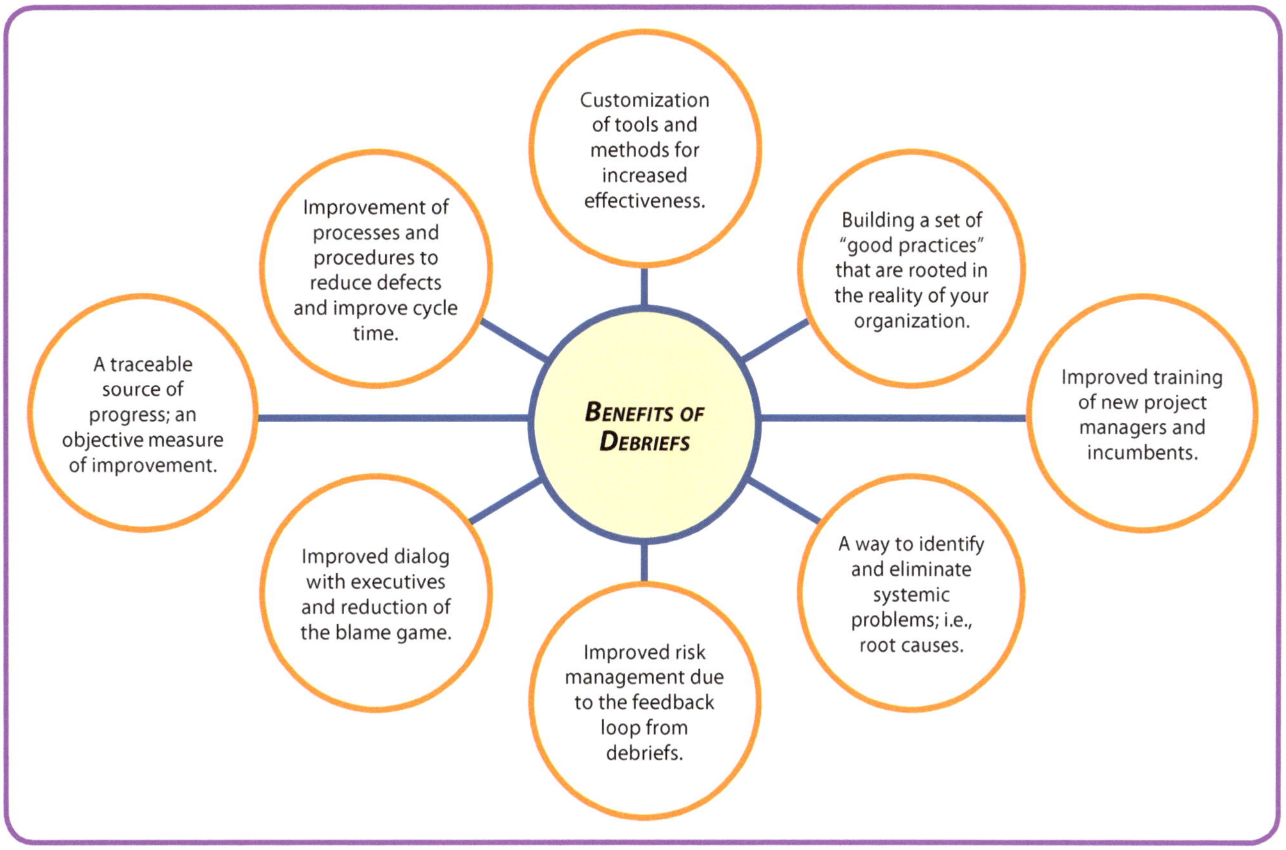

Phase 4: Close Out the Project

The Initial Questions

Before you call a meeting to debrief a project, there are three initial questions that must be answered:

1. Why are we debriefing this particular project?

   In order to justify the investment of time and effort, you must have a strategic reason to debrief a particular project — a criteria that, if met, flags a project as worth debriefing. We feel that not all projects need to be debriefed; rather, careful selection of the very few critical ones is more productive.

2. When will we do the debrief? (At what time during or after the project?)

   The timing of the debrief is critical so it can yield the right type of information. Debriefs are not exclusively done at the end of a project.

3. What will be the focus — the topic or emphasis — of the debrief?

   The answer to is connected to first, the reason for doing the debrief. You need not debrief the entire project.

The Sad Truths

- Most companies spend considerable resources planning and executing very expensive projects, and little or nothing on evaluating the results or learning to improve their performance, so they continue making the same mistakes.
- Few companies examine their completed projects in any depth.
- When they do perform debriefs, they are superficial and not documented.
- Employees participating in debriefs fear that there are hidden agendas and other nefarious intents on the part of the company.
- Few companies actually take action as a result of debriefs. There is no framework to ensure a closed-loop process.

This is not to discourage you from undertaking debriefs; rather, this is to bolster your resolve to ensure your debriefs are successful. Debriefs are successful when they result is visible productivity improvement.

You can take specific steps to remedy the five items listed above. Work with you team and your managers to ensure measurable results.

Resistance to Debriefs

In our research and experience we have found that organizations perceive debriefs as a very positive activity. Which begs the question: Why do so few organizations actually perform debriefs? We have found three critical reasons that keep organizations from undertaking debriefs. The intent of this section is to alert you to the possible obstacles so you can take the appropriate corrective action.

1. Fear of Punishment or Retribution

Regrettably, some debriefs deteriorate into finger-pointing and other forms of complaining. Worse yet, in rare cases the information is used to punish employees. This is wrong and must not take place. It stifles participation and ruins morale.

What to Do:

- When you schedule a debrief you must reassure everyone that no one will be punished or in any way scolded for speaking freely. This is called creating a safe environment. Obviously, you must adhere to this commitment. It is the right thing to do, and it will build your credibility for future debriefs.
- Be sure to talk to all managers and supervisors and obtain their commitment that they will only use the information for improving the productivity of the organization and not as a tool for punishment.
- During the meeting, allow a period of time for complaining and whining. It is human. It allows people to vent. Once this is done, you need to refocus the team on the task at hand — i.e., the debrief.

- During the meeting, remind the team members that it is safe to speak freely and that no negative action will take place. Then live up to your promise!
- In rare cases, you might want to simply interview individual team members for their input. For example, two of you take the responsibility of interviewing the members of the team and writing a final report with findings and recommendations. An alternative is to hire an outside facilitator or management consultant to perform the interviews and write the action plan.
- Another alternative is to hire an outside facilitator to help you with the entire debrief. The advantages are that the facilitator has the time to organize and ensure all preparations are done and someone neutral from outside the organization will facilitate the meeting.

2. Past Failures of Similar Efforts

This happens when previous debriefs or lessons-learned meetings did not produce any visible results. The organization is jaded or skeptical about the probability of success.

What to Do:

- Ensure that you are committed to take action on the input from the debrief. If you're not, then don't have the debrief!
- Develop an action plan with detailed information about what needs to be done, who is going to do it and when it will be done. You must exercise your leadership by tracking the implementation of the action plan.
- Ensure that managers and supervisors will allow employees to take some time to complete the action items from the plan.
- Communicate to all team members every time there are visible results from the action plan. This will encourage them to participate in future meetings.

3. Too Busy Working on the Next Project

In most organizations there is urgency to start on the next project, thus limiting the availability of people to complete the action items from the plan let alone attend the debrief meeting.

What to Do:

- From the start of your project, build in ample time for the debriefs. That is, include the debrief as part of the scope of the project and, more importantly, in the schedule and resource plan.
- Ensure that managers and supervisors will allow employees to take some time to complete the action items from the plan.
- Do not debrief every project — rather, carefully select the few that are likely to yield the most benefit.

Timing of Debriefs

Traditionally, debriefs are thought to be done at the end of a development project, but we feel there are several other opportunities to initiate them. We strongly suggest that you don't think of debriefs as something you do only at the end.

The timing for debriefs is critical to obtaining the maximum benefits. You need to use your judgment and decide the appropriate time and reason for debriefs. Remember, you do not debrief every development project — you must carefully select your targets.

The timing for your debriefs varies depending on your objectives. Why do you need the information? When do you need the information? And what are you going to do with it?

Phase 4: Close Out the Project

Opportunities to Perform Debriefs During the Product Development Cycle

The above diagram illustrates a generic New Product Development (NPD) process and the opportunities for debriefs. The objectives you set for your debriefs will determine their timing and number.

Detailed Timeline of Opportunities

Type of Debrief	Reason	Timing of Debrief
1. Pre-Birth	• Incorporate learning from previous projects. • Team building.	• Before or at the time of your project kick-off.
2. Interim or Progress Debrief *During key dates in the life of your project such as major milestones.*	• Impact change throughout the life of the project. • Early corrective action. • Improve communication.	• Planned times during the life of your project — use your judgment.
3. Team Handoff *Your project is transferred to another organization or location.*	• Inform new team coming on board. • Formalize the handoff.	• Major milestones. • When transferring to an outsource resource. • When transferring to manufacturing. • When transferring off-shore.
4. Customer Engagement. *Critical points in your project when considerable customer involvement is required such as in contract manufacturing or custom software.*	• Early feedback from your customers on a new product.	• During development. • After the new product has been announced. • As part of the acceptance process.
5. Crisis Recovery *Very useful to objectively diagnose the reason for failure and determine corrective action.*	• Solve current problem. • Recover from a major setback. • Eliminate repeating the same errors in the future.	• When the stuff hits the fan! • As soon as the crisis is identified. • At the end of the crisis period.
6. End of Project	• Evaluate the entire project. • Closure for the team. • Benefit future projects. • Inform management of your accomplishments.	• At the end of your project. • Shortly after the end of your project.

For comments, questions or additional copies please e-mail the author: flexpm@rapidinnovation.com.

Possible Focuses for Debriefs

See Also: Debrief Planning Tool by Focus

The following list illustrates some of the topics you might want to focus on. It's not a comprehensive list. You and your team can define additional focusses that are of special interest to you. Collaborate with your team to find the focusses that are critical to your future success.

Generally, debriefs are done at the project level, where the entire project is evaluated. While this is acceptable, it is more productive to focus your debrief on a specific area. Naturally, you invite individuals who can make a contribution to the chosen focus.

Focus (Topic)	Examples
1. Communications	• The effectiveness of communication within your team. • Also, with distant suppliers or contractors. • Communication with executives and managers. • Communication with your customers.
2. Planning	• The effectiveness of your planning efforts, tools and methods. • The collaboration and communication during the planning process. • Usefulness of your WBS. • Clarity of objectives.
3. Scheduling	• Focus on the duration of events, calendars. • Gannt charts. • Communication of the scheduling to all stakeholders. • Slippage and missed milestones.
4. Design Process	• The product design process. • Requirements management. • Tools and methods. • Interaction with vendors, contractors, consultants and suppliers. • Design review process. • Engineering change notices process.
5. Training and Coaching	• Effectiveness of the training associated with the project. • Overall training and competence building. • Coaching and mentoring.
6. Execution	• Ability to get things done when they are due. • Obstacles and barriers to get things done. • Dealing with breakdowns.
7. Reaction to Crisis	• Corrective action in view of a breakdown. • Collaboration to address an issue. • Avoidance of guilt and blame.
8. Risk Management	• Schedule accuracy due to proper risk management. • Ability to plan and monitor risks. • Dealing with risks that become reality.

Phase 4: Close Out the Project

Focus (Topic)	Examples
9. Teamwork and Collaboration	• Communication. • Recognition and celebrations. • Ability to communicate the common purpose. • Commitment to the success of the project and the team.
10. Resources (Human)	• Conflicts and obstacles. • Number of people in the project (resource allocation). • Skills, knowledge and abilities. • Intercultural communication and diversity. • Celebrations, recognition and ceremonies.
11. Leadership and Management	• Effectiveness of leadership. • Ability to manage the projects or the processes within the project. • Dealing with stress and pressure. • Accountability and consequences.
12. Budgets and Expenditures	• Planning. • Accuracy of budgets. • Fiduciary responsibility with vendors, suppliers, contractors and consultants.
13. Forecasting Accuracy	• Ability to develop accurate schedules. • Forecasting budgets, cost overruns.
14. Roles and Responsibilities	• Clarity of roles and responsibilities. • Clear communication when changes were made, • Clarity of roles and responsibilities with vendors and suppliers.
15. Project Closure	• Effectiveness of debriefs. • Residual design issues. • Documentation management. • Teamwork
16. Interaction with Customers	• Customer satisfaction. • Communication with customers. • Clarity of requirements and requirements process. • Customer complaints.
17. Vendors Contractors and Consultants	• Roles and responsibilities. • Costs and estimate accuracy. • Teamwork. • Obstacles, complaints and barriers.
18. Supply Chain	• Communication effectiveness. • Financial results. • Speed. • Defects.

Phase 4: Close Out the Project

Additional Information on Focusses

Some organizations select a set of standard focusses to debrief. This is acceptable, although there is a risk of repeating to the point where no additional significant information can be obtained. This is known as the point of diminishing return. Be careful not to fall into the trap of creating a formality or a perfunctory exercise with little or no value.

Your debrief strategy must be dynamic and able to change as your environment changes. Keep in mind also that every development project is different, so it makes little sense to become locked into a single approach for debriefs.

The same can be said about checklists, which can be useful, but can also lull you into a routine with little new information or insights. Checklists are used to ensure that all appropriate focusses are addressed. The risk is that after a few times it may become simply a way to pass the time. The intent of a debrief is to dialog, discover and address issues and opportunities, not to read a list.

Four Critical Decisions

There are four specific challenges to consider when planning, running and following up on a debrief. Work with your team and develop strategies and tactics to address them.

1. **WHICH PROJECT:** Which project to debrief?
2. **WHICH IMPROVEMENTS:** Which specific improvements will you implement?
3. **HOW TO DEPLOY:** How will you effectively deploy the needed changes to the organization?
4. **HOW TO SUSTAIN:** What is your strategy to make the needed changes a permanent part of the culture of your organization?

First Critical Decision: Which Project to Debrief?

The first challenge is to decide which projects you should debrief. Quality trumps quantity! Your organization has limited capacity to assimilate new information. So it's best to carefully select which improvements you will actually deploy.

We suggest that you start with a small number of projects per year — perhaps three to four debriefs per year or 10 to 20 percent of your projects. If in doubt as to how many, try two to three per year to start, then increase the number based on your ability to implement innovations resulting from them. It's best to do fewer due to variables such as the workload of your team or organization, the nature of your business and the overall number of projects that you complete per year.

It is not necessary to debrief all projects. Who has the time and capacity to process all the information? Instead, carefully develop a set of criteria that will help you maintain consistency in your choices.

Examples of Criteria for Selecting which Projects to Debrief

- Projects that were strategically important to your organization.
- Projects that failed to meet the documented objectives.
- Projects that were exceedingly successful.
- Projects that were particularly complex.
- Projects that were not traditional or typical — i.e., different from the majority of projects you lead.

Customize this list and generate your own priorities. Collaborate with your team and managers to develop consensus around a few criteria that will allow you to select only high-leverage projects to debrief.

Phase 4: Close Out the Project

Example: Selecting the Right Project to Debrief

The template below is an example of a simple tool you can use to select the right project. You can further develop the template to track, prioritize and optimize the results from debriefs.

Project Candidate	Business Case	Focus (Topic)	Plan Leader	Target Date	Deployment Leader
Selection of projects is based on a clear criteria, which is applied to all candidates.	The business rationale that justifies the expense in resources to perform the debrief.	The specific area of focus for the debrief. The desired outcome will dictate the focus.	The name of the person responsible for leading the planning and execution of the debrief.	The intended date when the main session of the debrief is planned.	The name of the person who is responsible to lead the implementation of the results of the debrief.
"Eagle Eye" platform	The investment in developing the new platform was over 10 million dollars. We must understand the effectiveness of the investment during the development cycle.	We will focus on our relationship with our vendors as 40% of the new platform was outsourced. We will also focus on the use of the new CAD software.	George Smith, VP of Supply Chain will lead the debrief process.	We will debrief two weeks after the product is announced to the market.	Mary Carter will be responsible for ensuring the deployment of the finding and continuous improvement.
"Columbia" (new product)	This project will not be debriefed as it is a simple improvement to an existing product.				
"NewWave" (new product)	This is our first product where more than 50% of the product is being outsourced to vendors across Asia. We must identify areas of improvement in our supply chain.	The focus will be in three areas: • Contract negotiation and execution. • Technical documentation. • Schedule management.	Lani Lui will lead the debrief; he is the project manager.	We will debrief twice. • At the mid-point of the project. • After the product has been rolled out to the market.	No person has been identified for deployment. For now our Quality Manager, Miguel Martinez, will monitor and make recommendations.

Second Critical Decision: Which Specific Improvements Will You Implement?

The second challenge is selecting the material from the debrief that you will actually deploy. Every debrief will yield a considerable amount of information, but only a fraction of it is worth deploying.

Here again, the development of selection criteria is important. An important step is prioritizing the information collected from your debriefs. As we outline in the process map for debriefs, every debrief should result in an action plan. In addition, every debrief will yield raw data — that is, information that is important but has yet to be ranked and sorted.

Processing all the information from your debriefs is a critical step. Processing means sorting by relevance to the organization, audience or constituency. For example, some information may be important to your marketing organization but not relevant to your finance group.

Examples of Prioritizing the Information

- Important and urgent. This is a two-variable criteria. The decision is made first by how important the improvement is to your organization and then by how urgent it is to deploy the specific improvement.
- Ability to leverage to improve productivity — i.e., which information will deliver the most performance improvement?
- By function — e.g., marketing, engineering, manufacturing, etc.
- Relevance to the executives of your organization. Allow the Executives of the company to prioritized based on the business objectives of the organization.
- Strategic information versus tactical.

Third Critical Decision: How Will You Effectively Deploy the Needed Changes to the Organization?

You must now select the method that you will use to deploy the information. What will be your strategy for ensuring that actual change occurs. In chaotic organizations it's always best to choose at least two ways to deploy. For example, through training and the quality manual. Or, through a web page and special presentations.

This is the most important step. It's the payoff for all the hard work. Consequently, you must dedicate considerable focus and effort. Failure to effectively deploy will render that debrief useless. Proper deployment will result in visible improvement in your organization's productivity.

Fourth Critical Decision: What is Your Strategy to Make the Needed Changes a Permanent Part of the Culture of Your Organization?

Finally, how will you ensure that change is long-lasting? So, you've decided how the valuable information will be deployed. But you are not done yet! The most often-forgotten step is tracking the deployment and providing enough support to ensure that the change A) does takes place and B) is long-lasting.

We suggest that you first form a Core Team to help you. This is a small group of people who volunteer to help you because they believe that the specific change you propose is important to the organizations. As a Core Team, hold a meeting to develop your action plan to sustain the change. Remember that at this point the change has been deployed (see the previous Critical Question).
Something else that will help sustain a change is to deploy a very simple metric to track progress. You can also request support from your manager and from the Human Resources personnel.

Essential

"Failure to effectively deploy will render that debrief useless. Proper deployment will result in visible improvement in your organization's productivity."

Phase 4: Close Out the Project

Key Questions to Consider

Will We Need a Facilitator for Debriefs?

In a Word, Yes

A facilitator is a neutral person who has the skills to lead your team in a productive search. A facilitator can create a safe environment, reassuring your team members that the intent is to improve performance, not to punish anyone. A facilitator focuses on the debriefing process, ensuring that your team is productive. Finally, a facilitator has the skills to expedite the process and focus your team on the right thing at the right time.

You can hire an outside facilitator. Another alternative is to ask your HR department to help. Or you could invite another project manager to facilitate your meeting. Perhaps a manager from a different organization might fit the bill. Do your best to secure a facilitator because it makes a considerable difference.

What About Multi-Site Projects

Product development for many companies takes place in many locations around the world, and this practice is growing. When separated by distance, language and culture the need for debriefs becomes more important. We strongly encourage you to do more frequent debriefs if your development team is scattered around the world.

The purpose of a multi-site debrief is the same as with a co-located one: Improve the future performance of your organization. The preparations for a multi-site debrief vary based on the distance, language and culture. To reduce travel costs, add debriefs to the last project meeting, which is already budgeted. Furthermore, there may be reasons that warrant having the team travel for a debrief — for example, a project that was exceptionally successful, one that failed or a project that represented a significant new strategy for your company.

Using Video Conference to Do a Debrief

If you cannot meet face-to-face, then you must prepare the logistics carefully. The next best thing to a face-to-face meeting is a video conference, as it allows for some interaction. The process of selecting which project to debrief and what part of the project to debrief is the same as with any other. The same can be said for what you will do after the debrief. Here are some steps to take initiating the process:

1. Start by selecting a volunteer in each of the locations who will serve as the "point person"; that is, your representative in that site. He or she will help with logistics, preparations and follow up after the meeting.
2. Schedule the meeting and the video facilities.
3. Send clear notices and talk with each team member — preferably by phone rather than email — so they understand when and where to meet.
4. Prepare the team by providing the right information prior to the debrief — for example, the objectives of the debrief.
5. Ask each team member to bring to the video conference a list of two things that worked particularly well during the project and two things that must be improved.
6. Use the same process to do the debrief taking into account the use of video. You want to create as much interaction as possible.
7. You need to have someone capture the information electronically, as you may not have the luxury of yellow stickies. You can use a spreadsheet to capture and prioritize the feedback. Prioritization may be done after the video conference to save time and money.

Be mindful of cultural and language differences. Some cultures are not comfortable with anything that appears to be negative criticism. Coach team members that the debrief is about the project, and not about any individual. Encourage them to write down the information prior to the meeting to remove some of the stress.

Phase 4: Close Out the Project

Language represents another challenge. Many people for whom English is a second language tend to read better than speak. This is another situation where asking them to write their input prior to the video conference is helpful.

You may want someone local to collect the information and present it during the meeting. This way other team members don't feel put on the spot.

The Role and Responsibilities of Managers and Executives

- Managers and executives should avoid attending the debrief meetings. Allow the owners of the problems to solve them. Also, too many leaders may result in stifling the flow of ideas. Finally, managers and executives may tend to intimidate due to fear of punishment.
- Managers and executives must be ready to provide support, funding and encouragement once the solutions have been identified by the team in the action plan.
- Managers must avoid any blaming or guilt during any stage of a debrief, which would inevitably have a negative impact during future debriefs to the point of rendering them useless. All debriefs are guilt-free! Use other means to appraise performance.
- Celebrate the genuine improvement in productivity due to debriefs. Remember that every dollar saved though productivity goes directly to the bottom line!

Getting to the Root of Things

During your debriefs, you will receive a set of symptoms — for example, "The document tracking software did not work." Naturally the question begs, "Why did the document tracking software not work?"

You need to be familiar with methods of root cause analysis such as the Fishbone Diagram or the Affinity Diagram. This way you can help your team identify the root of the problem, which will enable you to fix things once and for all. Additionally, The 5 Whys is a simple way to facilitate identifying the root cause of problems.

The moral of this story is to beware not to fix symptoms, but instead fix the cause of the problem.

Essential

"The moral of this story is to beware not to fix symptoms, but instead fix the cause of the problem."

See Also **Ways to Prioritize**

The Power of the Lowly Yellow Stickies

These little pieces of paper are tremendously helpful in making your debrief more manageable and productive. Ask your team members to record their input on yellow stickies; for example, Post-It Notes® (3M brand, model #655 3x5). This creates a documented record that can later be used to organize and prioritize the information.

Your facilitator will be familiar with several ways to use the yellow stickies to make decisions, organize and prioritize all the information from your debrief. While we're on pieces of paper, the use of flip charts (easels with large pads of paper) is also very important for debriefs. Make certain that you have an ample supply of these valuable aids.

The Protocol to Describe Debrief Input

It saves considerable time and improves the overall quality of your debrief to establish and enforce a protocol to describe the input from your team. When you start your meeting, describe the protocol, show a few examples and tell your team to start using it. During the meeting you will probably have to remind them to continue to use the protocol. You need descriptions that are clear. This debrief may be your best opportunity to obtain this input, and you might not get another opportunity.

Phase 4: Close Out the Project

Examples: Effectively Describe Issues and Opportunities

See Also: Yellow Stickies Protocol

The Protocol

Description of the issue or opportunity.	Describes what went wrong or what went right. (No names of individuals!)
Description of the possible root cause.	Describes the root cause of the problem. If you don't know it, then leave this section blank.
Description of the impact, damage or improvement.	Describes the impact, damage or opportunity. Remember, debriefs are also about things that went well and should be repeated.
	Author's Name

Example 1

Description of the issue or opportunity.	The Operators Manuals were not delivered on time
Description of the possible root cause.	due to a delay at the printing plant in India,
Description of the impact, damage or improvement.	which forced us to ship incomplete and poorly printed Operators Manuals for the first 1500 units.
	M. Change

Example 2

Description of the issue or opportunity.	The first prototypes were delivered on time, were faithful to the design and had excellent detail.
Description of the possible root cause.	We used a new supplier to try a new Rapid Prototyping machine.
Description of the impact, damage or improvement.	The speed and quality saved us at least two weeks in the schedule and provided a higher quality prototype.
	L. Medina

Rules and Strategies for Debriefs

Strategic Rules

1. The key question to ask before you do any debrief: What is its intent? You need a specific, documented purpose that shows why this debrief is important, why this focus for this project is worth the effort.
2. Before any debrief, ensure that there is a willingness to adapt and improve in your organization. Every debrief should result in some type of change.
3. All debriefs should reach specific, documented findings and culminate with an action plan or roadmap for improvement.
4. A debrief should be about how teams and organizations worked, not how individuals worked.
5. Debriefs are always guilt free — they are not performance reviews, and no one should be punished or reprimanded.
6. Not all projects need to be debriefed. Carefully select those that might yield considerable leverage.
7. All debriefs are done to create the capacity to improve future performance.

Tactical Rules

1. Debriefs should be planned and carefully choreographed to ensure quality input — they are not casual events.
2. The best tool for debriefs is yellow stickies (Post-It Notes®).
3. Debriefs should allow ample time for dialog and discussion — generally two to three hours for interim debriefs and a full day for end-of-project.
4. A debrief should always be a positive experience. Without open communication and common intentions, the effectiveness of the debrief will be diminished.
5. Debriefs can be held throughout the project, not just at the end.
6. Debriefs should include all team members whenever possible. It is the perception of the entire team that provides the insightfulness you desire.

Key Questions Before You Conduct Your First Debrief

1. Are the Executives of your organization in support of debriefs?
2. Do your team members have the training needed for successful debriefs?
3. Do you have a long-term plan to sustain debriefs for the long run?
4. Do you have the tenacity to stay with it?
5. Do you have a plan to demonstrate the value of debriefs once you start?
6. Do you know what you will do with the information from a debrief in order to improve productivity?

See Also Deployment

Phase 4: Close Out the Project

High Level Process: Debriefs

PLAN

1
- Which project will we debrief?
- Why are we doing this particular debrief?

2
- Who will participate?
- How and when will we perform the debrief?

DIALOG

3
- Brainstorm.
- Capture.
- Prioritize.
- Document.

ACT

4
- Action plan.

5
- Deploy.

6
- Ensure long-lasting change.

The above diagram shows a high-level process for debriefs. Note that it all starts with identifying the reason for your debrief. It is also important to know if you will have the means to implement the changes. Your team members and the rest of the organization will be disappointed if they do a debrief only to find out that none of you have the time, resources or support to implement the changes.

Detailed Process: Debriefs

WHERE YOU'RE AT

At this point you have used the criteria for selecting a project to debrief. Not all projects need to be debriefed — you must have a clear reason for selection.

The assumption is made that the project you have selected meets the criteria.

See Also Debrief Planning Tool

1

SET OBJECTIVES

Decide what you're going to debrief — for example, the process, the communication, leadership, flow of documentation or change notices. Or do a general debrief, open to all issues.

Decide what you will do with the information once you have it — i.e., how will you ensure that the input translates into action and eventual improvement.

DO NOT DO A DEBRIEF WITHOUT THIS STEP!

2

DOCUMENT YOUR PLAN

Use the Debrief Planning tool to help you prepare.

CONTINUED

Phase 4: Close Out the Project

CONT'D

Important

Ensure that you invite the participants. Inviting is much more than sending an email.

You need to talk to each person, and ensure that he or she has your debrief on the calendar. Reassure everyone that their input is valuable.

You might want to ask the managers, when they hold their staff meetings, to remind people to attend your debrief.

3

Schedule the Debrief Meeting

A good debrief should last at least three hours to allow for proper discussion, prioritization and action planning. For large and complex programs, a debrief can take as long as a full day, and in some cases longer.

Decide whom you need to invite given the objectives of the debrief. It's better to be inclusive rather than exclusive.

Collect relevant documents that can be used to help in the debriefing process. Make copies for everyone as appropriate.

Create an agenda — for example:

- Set the stage.
- Tell participants what will be done with the information.
- Review the history of the program to refresh people's memories.
- Open discussion using yellow stickies.
- Identify actionable items — i.e., items worth taking action on.
- Prioritize.
- Assign responsibilities for action.
- Agree on the next step, to ensure closure.

Important

Prior to the start of your debrief you developed an agenda. But sometimes even with an agenda it's difficult to get things started.

You might want to rebuild the path of your project — i.e., use a chronological approach.

You might want to use the functional approach — i.e., talk about the marketing, engineering or financial aspects of your project.

Another possibility is to start with the positive parts of your project — e.g., what worked?

You have the option of asking every participant to come prepared with two things that went particularly well and two other things that need improvement. This will help jump-start the conversation at your meeting.

Important

Select a large conference room to allow team members to break into small groups as needed.

Ensure that you have all the office supplies you will need; i.e., yellow stickies, black felt tip markers, flip chart easel and paper.

CONTINUED

Phase 4: Close Out the Project

Where You're At

At this point you have begun the debrief meeting.

Cont'd

4

Set the Tone for a Positive Event

Reassure the participants that no one will be punished, reprimanded or otherwise sanctioned as a result of this debrief. A debrief is not a performance review.

It is important that the entire team understand and agree that this meeting is not for punishing, but creating new opportunities, improving performance and reducing future stress — in short, it is a positive event.

Create a safe environment where people feel they can speak freely without fear of retribution.

Use yellow stickies to get the information out. Many people are not too eager to speak in a larger group.

PEOPLE MUST FEEL THAT THEY CAN TRUST YOU AND THE PROCESS!

Important

Avoid the blame game. Stay away from "right" or "wrong;" rather, focus on "worked" or "didn't work." Also, start by listing achievements. This will reduce stress.

Don't use people's names; rather, talk about the problems, issues and breakdowns.

Ask, "What should we do differently next time?" This will create a forward-looking environment and reduce the tendency to look back in blame.

5

Review the Project

Start with a quick review of the project. This is an opportunity to share the materials you've prepared; for example, the schedule and other documentation.

This is also a good time to remind your team of the specific focus; for example, communication, supply chain or technical development.

Important

Create a Parking Lot

This is a handy tool to capture information that is important, yet not relevant to the focus of your debrief.

Take a sheet of flip chart paper, label it "Parking Lot," and adhere it to the wall.

Encourage your team to add yellow stickies as the meeting progresses.

The Parking Lot will reduce the number of distractions. Plus, it will help you capture valuable information for later use.

6

Facilitate Brainstorming and Dialog About Ways to Improve

Use yellow stickies to document all input.

Some questions that might create a lively discussion:

- What worked?
- What didn't work?
- What should never be repeated?
- What should be emulated in future programs?
- What tools worked well? What tools did not work?
- What training was useful?
- What additional training is needed for future programs?
- How effective was the communication loop?
- What abilities or skills need to be improved?
- What created the most distraction? And what will we do differently next time?
- What one thing slowed down the program the most? And what will we do differently next time?

Continued

Phase 4: Close Out the Project

Where You're At

At this point several hours may have passed.

You will have collected a considerable number of yellow stickies. All of this is raw data as — it is not prioritized nor organized in any fashion.

It is clear that no more valuable input can be collected.

Cont'd

7

Organize and Prioritizes Necessary Improvements

Once you have covered your agenda, it's time to organize and then prioritize the input using consensus.

You must perform this step before declaring the session over. It is very important!

There are several ways to approach this critical step. For example, use the affinity diagram or story-boarding approaches. You might want to bring in a facilitator who is experienced in the use of these tools.

Prioritize all the input using a clear and simple set of criteria. For example, urgent and important, most needed for the next program or highest overall leverage.

See Also

Critical Tools

8

Develop Message to Management

At the end of the discussion (brainstorming), ask the team to develop a message to management: What one thing does the team wish management to know about this program? For example, perhaps it is something that needs to be improved but is beyond the responsibility of your team.

This is a separate document or presentation you and your team develop for your managers and executives. It is important because it engages them in the debriefing process. Plus, it's your opportunity to ask for support and resources to implement needed changes.

Your message should be clear but not confrontational. Avoid blaming management; rather, present your information as an opportunity and in a positive vein.

This message can be delivered to your management by a small group from your team, or you can do it yourself. Another possibility is to ask a willing manager to deliver the message during an appropriate staff meeting.

This is an optional step but one we highly recommend. Your executives will welcome your suggestion. Moreover, management can be very helpful in ensuring change that is long-lasting.

Important!

Your Message to Management needs to include your triumphs and successes! Debriefs aren't just about what needs to be improved but also all the good things that took place. Ensure that your managers and executives know about your team's results and breakthroughs.

And it's worth repeating: Keep all your reports and presentations blame free! Performance issues and incompetence can be dealt in separate forums — your HR people can help.

IMPORTANT NOTE: "Incompetence" simply means someone didn't have the skills, knowledge or ability to perform a job. It doesn't mean he or she is a bad person. We recognize that the term "incompetence" in most contexts has a negative connotation. We are using the term as neutral, and we expect you to use it the same way.

We cannot omit the statement, but we can rephrase it: "lack of competence," "lack of fitness" or "lack of abilities."

Important!

This is a good opportunity to also review your Parking Lot and extract relevant information in your Message to Management.

Continued

Phase 4: Close Out the Project

CONT'D

9

Conclude the Debrief Meeting

At the end of your debrief meeting you will have:

- A documented and prioritized list of improvements, be it "do more of" or "do less of."
- A list of ad hoc, tactical improvements with clear delivery dates and responsibilities. These are simple improvements that do not require much effort but can deliver much value.
- A Message to Management document.
- An archive of important information from your meeting, beyond the list of improvements.

See Also
A Well Documented Action Plan

10

Create an Action Plan

During the debrief, you organized and then prioritized the list of suggested improvements to be made. Now you need to translate this list into a roadmap — i.e., the action steps needed over a timeline.

As with any action plan, it must document clearly what needs to be improved, who is responsible for ensuring the improvement and when the improvement will take place. Otherwise, there is a high probability that all the effort will be wasted.

Key points:

- It may require a subsequent meeting of key players.
- Make certain that the plan is developed in collaboration with your team members.
- Distribute segments to the appropriate stakeholders — i.e., pass out the action items from the plan, but do not flood team members with information that is not relevant to them.
- Decide and document how you will know the changes have taken place.
- Don't forget the critical questions:
  - What must we do differently next time?
  - How will we make certain we change for next time?
- Ensure that all relevant information is documented and archived for your future use and for the use of other projects.
- You must create the closed loop to ensure the benefits of your efforts. Once all the action items are completed, you want to ensure that the good work done results in actual change — improvements — in the organization.

CONTINUED

Phase 4: Close Out the Project

Where You're At

At this point you have concluded your meeting and everyone has gone back to their jobs. It is now time for you to lead your action plan to results!

Cont'd

11

Implement the Action Plan

This is a case where dogged tenacity wins the day. You must stay on top of the action plan in order to ensure completion and, most of all, productivity improvement.

This is the roadmap that you and your team create to change things that are within your control. Generally, this is known as remedial action.

You will find areas of improvement in processes and systems beyond the control of your team. Ensure that your action plan differentiates between the things that your team will improve and those that need the attention of others.

12

Deployment

This is not the same as the action plan. It is the process by which you let every one else in your organization, your company even, know what needs to be improved.

Also, this is the place where you communicate good practices so other organizations benefit from your findings.

This information must be communicated proactively. You might want to recruit others who participated in the debrief to help you accelerate change.

This will challenge your commitment to make a difference. It is up to you to track the progress, and lead the organization into taking action.

13

Knowledge Bank

Critical information from all of your debriefs should go into a single storage area, which is called the Knowledge Bank or Best Practices.

This makes it convenient for your entire organization to access the results from all of your debriefs.

This is not a trivial function, as the data base must be kept up to date, prioritized, sorted and, most of all, convenient to your organization. It must not be some obscure folder on some remote server.

14

Long-Term Improvement

Once you have developed your action plan and you have done the job of disseminating the information, you are still not finished.

Many of the improvements and good practices that you discover require a cultural change. That is, you must lead the organization to drop old habits and adopt new ones.

You need to work in concert with your managers, executives and HR department to ensure that change does takes place.

Learn and Improve (Kaizen)

Learn and continue to improve your process. Over time you and your team will develop the needed expertise to perform debriefs that make measurable difference in your entire company.

STOP

Phase 4: Close Out the Project

Create a Well-Documented Action Plan from a Debrief

Warning

The biggest challenge about debriefs is carrying out the action needed to improve productivity. It is relatively simple to schedule the debrief with your team; it is difficult to ensure that actual change happens as a result of the debrief.

Before you call a debrief, ensure that you have the time, resources and support to implement the changes that will be expected after the debrief itself.

To ensure change you first need to develop an action plan — a detailed document with clear descriptions of the action items, unambiguous appointments of who will be responsible for carrying out the action items and a timeline indicating when the action item will be completed.

Example: A Well-Documented Action Plan

Item	Background Info	Action Item	Responsibility	Timeline
1.	Key specifications were not documented, which caused three weeks delay in the "Eagle" project.	The product marketing managers will call a review of the specifications of the MRD (Marketing Requirements Document), starting with the next project and before the MRD is sent to engineering.	Lou Chin, Product Marketing Manger.	Lou will report progress to the Core Team no later than October 10, 20XX. Lou will complete the action item no later than October 21, 20XX.
2.	There was confusion about the schedule during the "Eagle" project.	Create a large schedule Gantt chart using a white board and yellow stickies so it is visible to everyone on the team.	Mary Lynch, Engineering Manager.	Mary will work with the engineering team to create the schedule by the start of the next project (Eagle 2, January 2, 20XX).

The required action is clearly stated to avoid confusion.

Responsibility is clear by writing the name and title of the individual.

The action plan is not complete until clear dates are included.

This column explains the background of the problem as it was discussed during the debrief. This is helpful for reference.

The above example shows a well-documented action plan from a debrief. Do not call a debrief meeting if you do not have the resources to ensure the completion of the action items. A debrief has value only when it results in visible improvement in the productivity of your organization.

Example: Debrief Planning Tool

To be filled out by the person responsible for organizing the upcoming debrief.

Item	Description
Project to Debrief	Project "Eagle" will be debriefed because the supply chain model failed and caused a six-month delay.
Brief Description	"Eagle" was a project to develop a new product platform to address the needs of the sheet metal industry, which represents a new market for our company.
Specific Focus	We will analyze the schedule and the supply chain to extract and prioritize improvements for future projects. We will also analyze our Quality manual and the supply chain management function.
Date, Time and Location	November 5, 2009 Conference room #210 We will start promptly at 9:00 a.m. and finish at 12:00 noon
Attendees	• Project manager • Product manager • Engineering manager • Manufacturing line manager • Supply chain manager • Base data manager • Logistics manger
Reference Materials and Documentation Needed	• The project schedule • A list of the major delays and their impact on the overall schedule • The list of suppliers in the supply chain • A "block diagram" of the "Eagle" product • A unit of "Eagle" disassembled to allow the examination of key components

Phase 4: Close Out the Project

How to Actualize Productivity Improvement

The Biggest Challenge of Them All

If there is a common problem with most if not all debriefs, it is the way the information is translated into productivity improvement (deployment and adoption). The intent of all involved in debriefs is to cause change, so they and their peers don't have to suffer through the same problems. Yet most debriefs result in little or no change.

This is not the fault of the people doing the debriefs. It is the lack of a clear process, and sometimes the lack of leadership, that causes the dead end. Consider that there are really two major stages to every debrief. One is obtaining and prioritizing the information. The second is the transfer of that information in a way that causes change.

We suggest you develop the process to deploy the information before any debrief takes place. This can be done by a small team, a group of project managers, a facilitator or your HR department.

Key Questions when Developing the Framework for Deployment

- Who will collect and archive all the information from all the debriefs?
- How will that information be categorized for easy retrieval and future use?
- How will the decision be made as to which improvements will be deployed throughout the organization?
- Who will make the above decision?
- What will be the process to deploy the information?
- What will be measured to ensure that the deployment is working?
- What will be measured to ensure that improvement is taking place as a result of the debriefs?

> Essential
>
> "Those who cannot learn from history are doomed to repeat it"
>
> — George Santayana

Who Can Help?

- Recruit your HR people to help develop an action plan to ensure progress.
- Create a team (Quality Team, Continuous Improvement Team, etc.), and charter it to carry out the change over a set period of time. These are small teams of three to five people who can implement the specific changes. These teams are generally made up members from your Core Team, but you can tap into other groups when it makes sense.
- Get the executives of your organization to commit to carrying our your recommendations. They can turn them into an "initiative" as part of their business or operational planning process.
- Some organizations have a quality manager who is responsible for ensuring continuous improvement.
- Form your own "Tiger Team" among the people who participated in your debrief, and help them carry out the needed changes.
- Make a commitment to drive some of the change yourself. You'd be surprised how much one individual can influence an entire organization.
- Work with your training organization to incorporate the changes into the training of new and current employees. Your training department can also incorporate changes in the documentation used for development.

Phase 4: Close Out the Project

Spreading the Knowledge

Once you've prioritized the information gleaned from the debriefs, it's essential to promote awareness and acceptance of the lessons learned. Don't assume your organization will change automatically or even quickly!

Below is a short list of how you can spread the knowledge. Based on your organization's culture and reporting structure, you and your team can find even more innovative ways to deploy information so it creates long-lasting change. The key question remains: What is the best way to ensure performance improvement?

Examples	Description
1. Training and Development	Continually update your training materials to reflect the improvements. Ensure that all new project managers are trained on the lessons learned, and run incumbent project managers through refresher training.
2. Brown Bag Sessions	Create a series of 60- or 90-minute sessions during lunch, breakfast or before or after work, facilitated by a project manager. Offer a meal or refreshments.
3. Good Practices Web Page	Develop a "good practices" (some companies call them "best practices") web page for your intranet, extranet or internet site. Make full use of the internet to keep everyone fully informed and trained, including those who are in other states or countries.
4. Community of Practice or User's Group	Start a group of people who volunteer to meet regularly on a particular topic because they have a passion for said topic. The organization provides support in the way of facilitation and resources.
5. Executive Retreats	Planning meetings, retreats and off-sites can include time to present the lessons learned. But you must be proactive and have much tenacity to prevail. Get your Management Mentor to schedule the presentation. Better yet, show them your leadership by scheduling it yourself.
6. Staff Meetings	Schedule a presentation for 15 minutes at the start of your staff meetings, and give it priority. Or dedicate one full staff meeting each quarter to discuss your debriefs. Be aggressive, and schedule a half-day staff meeting just on debriefs.
7. New Projects	At the start of all new projects, allocate 60 to 90 minutes to talk about lessons learned from previous projects or review good practices.
8. Required Reading	Document improvements, make them convenient for all to read and then make reading them a requirement — no excuses!
9. Tools, Templates and Guidelines	Incorporate performance improvements into tools, templates and guidelines. Constantly update your documentation to reflect the lessons learned.
10. Quality Manual	Incorporate performance improvements into your quality manual. Use Six Sigma, ISO-9000 or your quality system to ensure that changes are incorporated in the processes and SOPs.
11. Special Meeting	Have a special meeting after the results of a debrief have been documented. Dedicate quality time to discussing the recent findings, and invite managers. This is done before any documentation is published — do it ad hoc or extemporaneously.
12. Project Management Excellence Meeting	Once or twice a year hold a Project Management Excellence meeting. This can be a half-day or even a full day to share the improvement and engage the entire organization in continuous improvement.

Phase 4: Close Out the Project

Example: Debrief Planning Tool by Focus (Topic)

This is a comprehensive form to enable conversations. It may be too complex for some projects or programs, so we suggest that you customize this form.

Do not send this form to team members! This form is to help you facilitate the meeting, but the real value is the dialog.

1. Planning

   For example:

   - ❖ Were the goals, purpose and objectives clear to the team?
   - ❖ Was the plan for the project sufficiently clear to enable the team to deliver results?
   - ❖ Were the executives and managers available to clarify the goals of the project?
   - ❖ What steps will improve the planning of future projects?

2. Resources

   For example:

   - ❖ Were the resources properly planned?
   - ❖ Were the contractors, temps and consultants available when they were needed?

3. Scheduling

   For example:

   - ❖ Was the initial schedule realistic?
   - ❖ Were the project milestones helpful in creating progress?
   - ❖ What were the obstacles to meeting the schedule?

4. Performance of Project or Program Management

   For example:

   - ❖ How did the project plan work (WBS)?
   - ❖ How well was the project defined?
   - ❖ How well was it tracked?
   - ❖ How well was the plan communicated?
   - ❖ How well was the plan documented?
   - ❖ How widespread was the use of the project plan?
   - ❖ Was the plan realistic from the start?

Phase 4: Close Out the Project

5. R&D / Technology Management

For example:

- ❖ How was the project managed at the R&D level?
- ❖ How well did R&D communicate methods, technology, barriers and risks?
- ❖ Any breakdowns in R&D? Why?
- ❖ Were technical issues identified early enough?
- ❖ Were technical issues communicated early enough?
- ❖ Was there an effective method to address technical issues?
- ❖ Were estimates of time and cost on track?
- ❖ What interfered with R&D?
- ❖ What helped R&D?
- ❖ What were the lessons learned?

6. Communications

For example:

- ❖ Were all stakeholders informed?
- ❖ Did the team communicate effectively (the right information to the right team member at the right time)?
- ❖ Were remote team members properly integrated in the communication loop?
- ❖ How about team members in other countries (language, culture, timing, etc.)?
- ❖ Was the communication with managers and executives effective?
- ❖ What must be done next time to improve communication?
- ❖ What should we never do again when it comes to communication?

7. Managers and Executives

For example:

- ❖ Did your managers help the team be successful?
- ❖ Were decisions made on time?
- ❖ Were decisions properly communicated to the team?
- ❖ What should managers and executives do next time to improve overall performance?

8. Product

For example:

- ❖ Was the product delivered to the customers on time? Was it available to order on time?
- ❖ Was the introduction to the market place effective? As a public announcement? Through sales channels training? Through major accounts? Internationally?
- ❖ Was the promotional material available at the right time?
- ❖ Were the customers satisfied with the product rollout?

Phase 4: Close Out the Project

9. Marketing

For example:

- ❖ Was the product launch effective? Says who?
- ❖ Was the product proposal effective? Says who?
- ❖ Did the market message work? Says who?
- ❖ What slowed down the introduction?
- ❖ How was marketing handled in the venue of: Sales training? Promotional materials? Internet pages? Major accounts? International?

10. Manufacturing (In Software Development, use Documentation or QA)

For example:

- ❖ Was the team properly prepared to receive the product?
- ❖ Did the product meet the manufacturability targets?
- ❖ What was the root cause of any unexpected delay?
- ❖ Was the BOM (Bill of Materials) accurate?
- ❖ How about the supply chain? Did it work as it was planned?

11. Virtual or Distributed

For example:

- ❖ Were the tools used to work with distant team members appropriate?
- ❖ How about document management? Engineering documentation?
- ❖ Was email valuable? Why? If not, why not?
- ❖ How about project management in distant locations?

12. Service and Repair

For example:

- ❖ Was there an effective service plan?
- ❖ Was it on time?
- ❖ Was it communicated clearly?
- ❖ How about service training, service options, international service, spare parts, repair manuals, diagnosis, metrology and bud tracking?

13. General Questions

For example:

- ❖ What went exceptionally well?
- ❖ What went exceptionally poorly?
- ❖ What was the most important lesson we learned?
- ❖ What should we never do again?
- ❖ What should other projects remember about ours?

Critical Tools

Deployment of a New Tool

NEW TOPIC

Cascading a New Tool to the Organization

There is a difference between attending a seminar or workshop and causing your organization to embrace any or all of the tools you learned. Said another way, just because you and a number of your peers attend a seminar, that's no guarantee that what you learned will become common practice in your organization.

Deploying a tool takes additional work. It is a process, and requires planning, tenacity and leadership. Consider that you are not just enabling the adoption of a new tool or method, but actually changing old habits. In short, you will cause disruption and a break from the comfort zone currently in existence in your organization.

The Business Case

The return on investment (ROI) from any seminar or workshop only occurs after the class and when visible and measurable improvement in productivity is achieved. Some have the erroneous notion that attending a seminar is the objective; instead, attending is a means to and end. The real objective is to take what was learned and apply it to improve productivity.

Expect Resistance to Change

However, just because you and a number of your peers attend a seminar or workshop, that's no guarantee that what you learned will become common practice in your organization. Deploying a tool takes additional work. It's a process that requires planning, tenacity and leadership. Consider that you are not just enabling the adoption of a new tool or method, but actually changing old habits. In short, you will cause disruption and a break from the comfort zone currently in existence in your organization.

Expect resistance to change in overt and passive ways when you suggest the adoption of a new tool or method. You will hear things like "It won't work here," "We tried that, and it didn't work," "Our situation is different," and many similar statements. There are numerous resources about driving change in an organization to help you.

The Hesitaters

The initial push-back may actually be the expression of other feelings:

- **SEEKING ADDITIONAL INFORMATION:** Some in your team resist as a way to obtain additional information, as they feel fearful to deploy something they do not understand. It's the best way they know for doing fact-finding.
- **SEEKING REASSURANCE:** They feel unsafe; for example, they want to know that they will not get fired if the tool does not work the first time out.
- **TESTING:** They are simply trying to determine if this is for real and if you have a strong conviction that the tool will work.
- **GENUINE CONCERN:** Some may actually feel that the tool will not work or is not applicable.

The way to deal with all the above feelings is to first be aware that they are common, and then validate the actual source of the concern by providing additional information, reassurance, etc.

The Reactionaries

Some individuals are simply opposed to change, reactionaries who want to preserve the status quo. Occasionally you will run into someone who will refuse to change, refuse your new tool and steadfastly remain opposed. It is rare that you are able to change these individuals, and we don't have the time here to

Important! Much that has been written about driving change in an organization. When searching for books, seminars, articles and other resources use the terms "Managing Change" and "Resistance to Change."

Also, your human resources department may be able to help or offer additional resources.

delve deeply into this topic. We suggest that you do not become discouraged; rather, build a constituency of those who do want to change, and use it to overcome the resistance.

Food for Thought

Your strategy to overcome resistance must take into account that reluctance to change is a part of the human condition. The process to overcome resistance is all about people. Do not rely on templates, software programs, process maps, etc. Rather, seek to understand the situation, and then tap your leadership skills to drive change.

Here are a few things to keep in mind:

- First, set about understanding the current situation of those who need to adopt the new tool or method. Talk to them as a group and, more importantly, individually. Make an effort to listen with the intent of understanding.
- Be sure to consider WIFM (pronounced *wiffem*), which stands for, "What's in it for me?" We react positively when we understand our personal benefits. You must make the case for embracing the change such that every person on the team sees his or her benefits as part of the deal.
- Stop selling and start communicating! Our excitement for the new tool gets the best of us, and we start pushing the idea to the point where it becomes annoying. Instead, remain open to suggestions. Some will try to help you succeed by providing ideas for improvement — for example, a better way to use a template or a better way to deploy the tool. Be open to ideas, and engage those who provide them; do not reject them either directly or passively.
- Speak their language. Adapt your message to the audience. Speak in a way that is familiar to them. This will improve communication and avoid the use of new words.
- Start small. The intent of the first use of your new tool or method is to train and to develop the organization; thus, do not select a major project. Rather, find a simple project where the use of your new tool will shine and the risk of failure is very small.
- Finally, your tenacity and leadership will carry the day. Don't be disappointed with setbacks. In the end you're doing the right thing for your team and your company.

The Critical Questions

All of us need a certain amount of information before we commit to a tool, method or practice. It is not fair to expect people to jump at the opportunity if they don't have sufficient context, especially at the personal level.

Generally, we all want the following questions answered:

- Why are we doing this?

  (What is the relevance to my team, our company or our organization?)
- What would happen to us if we don't do it?

  (What would be the impact on or consequences to the company, organization or team?)
- Why are you asking me to do this?

  (Tell me why I should care. What's in for me?)
- What support can I expect from you?

  (Can I rely on you? What is your commitment to me?)
- What specifically do you expect me to do?

  (Tell me specifically how this impacts my job. What specific changes must I make?)

The above information must be communicated in the language and context of the listener. Avoid platitudes and irrelevant information as they tend to confuse and add uncertainty.

Deployment of a New Tool

Detailed Process: Deploying a New Tool

1

Choose the Tool

It's best to start small. Deploy a single tool to reduce the challenge and risk of failure.

Define clear criteria; for example:

- Ease of learning.
- Amount of leverage the tool will provide.

You can select any criteria, but you need to be consistent in your evaluation and prioritization.

Start recruiting supporters and allies who react positively to the new tool.

Important

Use the Planning Document in this section to ensure that you are well prepared for the deployment.

2

Identify Your Core Team

You will need a small group of your peers to help you deploy the tool.

Team members should be willing and able to spend the needed amount of time helping you with the task. Sometimes it may take up to four hours per week.

Make sure the team will be in place for at least three calendar months. It might take longer.

Important

Remember to secure the visible support of your organization's executives. Without it, you risk failure.

Meet with your executives, and show them the benefits and improvements that your new tool will bring to the organization.

If possible, secure a sponsor or a mentor — one of the executives who is willing to assist you in the deployment.

Seek support from others who are willing to help deploy.

3

Secure Support from Your Executives

It is prudent to identify one of the executives of your organization to sponsor your efforts.

Sponsors, also called Mentors, will provide moral support, help remove obstacles and add a measure of stature to your efforts.

The best sponsor is one who, like you, has a passion for the tool you are leading.

Important

As you select the pilot for your deployment, also ensure that you have training, templates and other needed support to help the team succeed.

Continued

Deployment of a New Tool

CONT'D

4 — IDENTIFY THE PILOT

This is the team, organization or group of employees that will apply the tool to a real project.

The results of this pilot will provide valuable information when deploying the tool to the entire organization.

Remember to obtain approval from the appropriate managers and supervisors.

Important: Equip the Pilot

It is critical that you provide coaching, encouragement and support to the pilot team. There is the risk of push-back and disappointment if the tool does not work as expected

It would be best if you were part of the pilot team. That way, you will be there throughout the process.

5 — CONDUCT THE PILOT

Ensure that you provide a high level of support to the organization such that the deployment of the new tool does not end in failure.

Provide all the training, tools, encouragement and other forms of support throughout the deployment of your pilot.

Important: What if the Pilot Fails?

Even the best of efforts may results in failure.

The first thing to do is to debrief and identify the root cause of the failure — without this step you cannot recover.

Avoid blaming, guilt or punishment as these things add no value to the recovery process.

Once you have the root cause and lessons learned, modify the process and try a new pilot.

6 — DEBRIEF THE PILOT

The topic of the debrief is to identify what worked and what did not so you can make appropriate improvements for the full deployment.

This debrief is done with your Core Team.

7 — PROCEED WITH FULL DEPLOYMENT

Deployment of a New Tool

Planning Checklist

Template

Deploying a tool is indeed a project. Consequently, ensure that you have a deployment plan, a schedule, budget, etc. Working with your Core Team, or with the peers who have agreed to help, you can develop a tactical plan that will ensure a successful deployment.

The list below is intended to help you formulate your own plan. You may need to add your own items to the list. Also note that the list below is in no particular sequence. Finally, the list is not rigid. Use your judgment and do what makes sense within your organization.

Planning Checklist

- ❑ Schedule for the deployment defined, documented and approved.
- ❑ Training materials for employees developed and documented.
- ❑ Training materials produced — hard and soft copies — available for use.
- ❑ Training materials placed in the intranet or web page.
- ❑ Trainers of the new tool identified and informed of their duties.
- ❑ Training of employees scheduled and employees notified.
- ❑ Training of managers and supervisors scheduled.
- ❑ Feedback mechanism in place.
- ❑ Metrics to measure the progress and eventual success of the deployment identified and ready to use.
- ❑ Coaching and encouragement for managers, employees and stakeholders defined and ready — including those who will do the coaching.
- ❑ Support available; for example, facilitators, coaches, tech support, assistants, etc.
- ❑ Obstacle removal in place.
- ❑ Templates, checklists and other tools ready and available.
- ❑ Information on our web page done.
- ❑ Processes that will be impacted by the deployment identified and corrected.
- ❑ Changes to our quality manual completed.
- ❑ Manuals, guides and other reference materials that will be needed by the users ready and available.
- ❑ Executives briefed and ready to fully support the deployment.
- ❑ Human resources (HR) informed.
- ❑ Training department informed.

Deployment of a New Tool

Planning Document

This template may help you secure assistance from your executives. It is also a good point to start the process of deployment. It may even become a standard template for deployment of future tools and methodologies. Modify this template to meet your specific objectives within your organization.

1. Tool or methodology to be deployed.
2. Objective of the tool deployment.
   - What will be the outcome once the tool has been successfully deployed?
   - Why are you deploying this tool?
3. Who is the expert on this tool?
   - Who is the internal person who knows the tool and is willing to provide guidance to others?
4. What documentation is available?
   - Models.
   - Software.
   - Books.
5. Are there any external resources available?
   - Consultants.
   - Instructors.
   - Facilitators.
6. What are the risks involved in the deployment?
7. How will we pilot the tool?
   - Which team, organization or group will be the first to apply the tool to actual work.
8. What will be our communication plan?
   - With executives of the organization.
   - With employees.
   - With the pilot team.
   - With other stakeholders.
9. What will we measure to assure us that the deployment has been successful?

Deployment of a New Tool

Tips for Successful Deployment

There are a myriad of things you must consider before you deploy a new tool in your organization.

Brainstorm with your team to find innovative ways to ensure the success. Your corporate culture, industry and reporting structure will help you determine the best way to embed the new tool.

Key Questions

- How can we make the new tool a habit in our organization?
- How do we turn the novelty of the tool into the normal way we do things?

10 Tips for Successful Deployment
1. START SMALL — PICK A SIMPLE PROJECT FIRST *The intent at this stage is to ensure that your team knows how to use the new tool.*
2. CONSIDER THE CULTURE OF THE ORGANIZATION *If your organization has not used this tool before, you need to lead your team to understand and embrace it. Some will push back, and you need a plan to deal with some skepticism and denial.*
3. PROVIDE THE RIGHT TRAINING TO THE RIGHT PEOPLE AT THE RIGHT TIME *Your team, and even some managers, may need training. You must provide the skills necessary to succeed.*
4. ENSURE CROSS-FUNCTIONAL PARTICIPATION *Most tools provide maximum benefit when people from several functions participate. Moreover, most projects involve people from several areas of the organization, and they must be included.*
5. PUT IN PLACE A PROACTIVE FEEDBACK MECHANISM *Enable your team to provide you with input on the new tool.*
6. BE AVAILABLE TO REMOVE OBSTACLES *Inevitably, your team will run into obstacles when implementing a new tool for the first time. Put in place a mechanism that will enable you to remove these obstacles; otherwise, the team may reject the tool altogether.*
7. DEVELOP AN ESCALATION STRATEGY *Some of the issues and challenges that will emerge in deploying your new tool will need the attention of your managers and other executives. Plan an escalation process before things come to a boil.*
8. MAKE INFORMATION ABUNDANTLY AVAILABLE *When first deploying a tool, your team, managers and executives will need information. Make certain that information is abundant and conveniently available to all.*
9. MAKE COACHING AVAILABLE *Your team members will need one-on-one assistance. Even if you provide training, individuals will have questions and other barriers that you must help remove.*
10. CELEBRATE *Recognize progress — celebrate small and large victories. Everyone needs reassurance that progress is being made.*

NEW TOPIC

Problem Solving Process

Introduction

Program managers spend considerable time solving a stream of problems due to the fluid nature of technology. Many of these challenges are straightforward and can be readily dispatched.

However, sometimes tough problems are under-managed. Organizations wind up drifting, unsure of the reasons for change, the severity of the problem or the path forward.

A disciplined problem solving approach should be applied for the most difficult crises that will shape the direction of the business.

General Rules for Solving Problems

- Do not underestimate the need to apply a formal process to solving a problem. Some in the team might dismiss it as trivial or something that can easily be resolved.
- Beware of the easy solutions. Many times the problem is a symptom of a much larger issue. There is always the desire to jump to the solution without the process to diagnose the real root of the problem. Here is where a formal process of solving problems plays a very important role.
- Widely communicate the solution. Frequently a problem has been solved, but the majority of the team does not know it. The process for communicating the disposition of a problem should be as rigorous as the process to solve it.

Overview of the Problem Solving Process

1. Identify the problem. Come to a common agreement on the nature and extent of the issue.
2. Generate criteria for a solution. Select which business measures will guide the decision process.
3. Generate alternative solutions and test them against the business criteria you have selected.
4. Conduct diligent reviews on key options. Prune the options to the few that merit more detailed analysis against the criteria.
5. Decide! Make a clear decision and then back that decision with a detailed implementation plan.
6. Communicate the solution to all stakeholders.
7. Learn from the process

Important

TIPS is the acronym for Theory of Inventive Problem Solving. TRIZ is the acronym for the same phrase in Russian.

TRIZ was first developed in 1946 by Genrich Altshuller and his colleagues in the former USSR and is now being developed and practiced throughout the world.

TRIZ research began with the hypothesis that there are universal principles of invention that are the basis for creative innovations that advance technology. Furthermore, if these principles could be identified and codified, they could be taught to people to make the process of invention more predictable. The research has proceeded in several stages over the last 50 years. More than two million patents have been examined, classified by their level of inventiveness and analyzed to look for principles of innovation. The three primary findings of this research are as follows:

Problems and solutions were repeated across industries and sciences.

Patterns of technical evolution were repeated across industries and sciences.

Innovations used scientific effects outside the field where they were developed.

For additional details visit: http://www.triz-journal.com.

Problem-Solving Process

Detailed Process: Problem Solving

1

Prepare for Your Problem Solving Meeting

Being well-prepared will make your meeting more successful and quicker:

- Gather all materials that will help you add clarity to the process; for example, technical documentation, reports and drawings.
- Select the attendees based on their ability to help solve the problem. Do not hesitate to invite people who are not directly involved with the problem but can add value to it. For example, experts on the topic and outside consultants.
- Ensure diversity of participants; the more points of view the better. For example, combine people who are involved in the problem with those who are not. People from other functions may also be helpful, as will some of your suppliers and vendors.
- Take care of all logistics; for example, reserve a venue that is large enough for your team to work.

2

Start Your Problem Solving Meeting

Share the perceptions of the problem: How does each team member see the problem? This is a critical step, as the term "problem" may mean different things to different people. Spend enough time in this definition phase — don't shortchange it.

Write the various perceptions on the flip chart. You can also use yellow stickies.

Clarify the definition of the problem by agreeing on a single wording of the problem.

Identify and document on a flip chart the expectations for a solution from your team members. As with the definition of the problem, the perception of a solution will vary. It is important to reconcile; i.e., find a common purpose.

Important

We have various ways to identify something that is not desirable; for example, we might call it a flaw, failure or fiasco.

Some of these terms carry a certain stigma or prejudice. Some cases imply punishment.

It is important that your team agrees on a term that is not politically loaded, but rather brings objectivity to your dialog. You cannot afford distractions.

3

Confirm Problem Statement

At this point you will have spent a considerable amount of time writing and re-writing the problem statement.

Write the problem statement one last time, as clear and readable as possible.

Validate with your team by asking everyone to agree or disagree with the statement. Does it truly represent the problem to be solved?

This is the most critical step of this process, the clear articulation of the problem, plus the support of the entire team.

You may need to spend additional time refining the problems statement, which would be time well-spent!

Do not proceed until there is clear alignment.

Important

The problem statement — the one you and your team have crafted — will most likely be the hardest step in the entire process.

Don't hesitate to rewrite the problem statement several times; it's worth the effort.

Spend all the time you need to ensure clarity and consensus among team members.

Do not proceed until everyone understands the statement of the problem.

Continued

Problem-Solving Process

CONT'D

4 PROFILE THE PLAYING FIELD

Analyze the problem based on the:

- **GIVENS:** Things that cannot be changed as part of the solution to the problem such as raising prices far in excess of what the customers are willing to pay.
- **FORCES AT WORK:** Those that are acting for and against the solution. Also known as Force Field Analysis.
- **MUST-HAVES:** Things that are non-negotiable or cannot be changed. Very similar to the givens, but with a focus on unacceptable trade-offs; for example, a reduction in product performance.
- **DESIRABLES:** Things that would be nice to have as part of your solution.

If appropriate, review the initial objectives of the project.

The outcome of this section is a single definition of the problem — the problem statement — and a set of criteria that bind the problem (a combination of steps 1 and 2).

5 DEFINE AND DOCUMENT SOLUTION CRITERIA

- Describe an effective solution (use the Yellow Sticky Protocol).
- Identify metrics to measure success of solution.
- Identify and list costs, customer needs and resources.
- Define the must-have criteria.

6 GENERATE ALTERNATIVE SOLUTIONS

- These are possible solutions, not necessarily real ones. This is a critical step in order to tap the creativity of your team.
- The rules of brainstorming apply, such as don't criticism and keep an open mind. This is not the time to eliminate or stifle ideas.
- Use a brainstorming tool like an Affinity Diagram.

Important!

We recommend using yellow stickies for the step of generating alternative solutions.

Ask your team to write down every possible solutions to the problem. Then ask your team to read each suggestion before posting every yellow sticky on the wall.

You can use the Affinity Process to organize and prioritize the yellow stickies.

See Also **Critical Tools**

7 EVALUATE ALTERNATIVES AGAINST THE CRITERIA

- Evaluate the proposed solutions (step 3) against the developed criteria (step 2).
- Retest if you have other alternatives to consider.
- Don't underestimate the impact of varying skill levels and cultural gaps.
- Develop a matrix and consider combining alternatives.

Note: You may need to do research in order to validate that the alternatives are viable and achievable. This may take time, even require another meeting. Some teams opt to simply select the alternative, assign responsibility and proceed without validation.

CONTINUED

Problem-Solving Process

CONT'D

5

CONFIRM YOUR SELECTION OF SOLUTIONS

- Ensure clear understanding and alignment behind the preferred alternatives and the criteria.
- Conduct due diligence on the top options.
- Complete the fact-finding.
- Document the key assumptions.
- Document risks.
- Check for unintended impacts.

6

DEVELOP YOUR ACTION PLAN

- Develop a detailed plan to verify or validate the assumptions.
- Develop a plan to address the identified risks and uncertainty.
- Establish ground rules for the implementation of the solution. For example:
  - How are you going to deal with changes?
  - How will you monitor progress? Will you track the metrics, milestones or compliance?
  - When and how will you decide that the problem has been solved?
  - Who will be responsible for implementation?
  - Who will be responsible for closure?
  - Who, when and how will you declare victory?
- Close any gaps concerning resources, skills and culture (a gap is a misalignment between "what it is" and "what it needs to be"; for example, you need more resources that you currently have).
- Communicate the decision to all stakeholders.
- This step is all about ensuring execution; that is, the implementation of a road map that will guarantee the solution of the problem. This is a high-risk area, as many solutions fail to be implemented due to the lack of a plan of action.

LEARN AND IMPROVE (KAIZEN)

Your risk management process will improve significantly if you constantly add the lessons you've learned.

- Debrief the team on the process used to solve the problem.
- Identify areas of improvement.

Problem Solving

Flexibility Matrix for the Problem Solving Method

Strengths	Weaknesses
A traceable path of problem to solution.	Need to watch for hidden boundary conditions.
Ensures a lot of input from everyone.	A solution will be found, but it may be politically incorrect.
Participative; creates ownership and encourages diverse solutions.	It involves hours of work, and it may be tedious.
Appropriate for difficult problems.	
A way to dispose of a problem once and for all.	
A disciplined way to identify the real causes of a problem, not just opinions and anecdotal information.	

Flexibility

- You can vary the size of the team depending on the complexity of the problem.
- You can create additional steps to suit the culture of your organization.
- You can do the process in two or three sessions to avoid fatigue.
- Include the documentation in the archive of your project in case the problem shows up again.

The Problems that Refuse to Die

With some regularity, there are problems and issues that have been solved, yet they persist and continue to distract the organization. This may be due to a lack of proper communication; specifically, the problem was solved but only a few in the organization know it. This can be addressed by a proactive, high-energy communication effort. For example, clearly announce the solution during review meetings and send notices out specifically on the conclusion of the issue. It helps to be emphatic.

The other reason for this type of problem is that some in the organization do not wish it to go away. Keeping the problem alive is one way to fight the chosen solution. One way to deal with this difficulty is to use the formal process outlined in this document, which allows you to cover all the possibilities and overtly deal with and dispose of the problem — this is called an intervention. Simply put, deal with the issue directly.

Other Tools for Solving Problems

There are several tools that provide a fine framework for solving a problem; for example, the Affinity Diagram, the Fishbone Diagram (Ishikawa Diagram), Karnaugh Maps (2X2 matrixes) and various cause-and-effect diagrams. Some of these are covered in this document, and the rest are readily available on the Web and in other handbooks and guides.

More advanced methodologies are also available to tackle very complex or esoteric problems; for example, FMEA and TRIZ. These methodologies are also documented on the Web.

Avail yourself of these tools. We suggest that you become proficient in at least one or two.

Karnaugh Maps: 2x2 Matrixes

Prioritize Items from a Selection of Many

Introduction

Karnaugh Maps, also known as Karnaugh Diagrams, have their origins in mathematics and are related to Venn Diagrams . They are especially useful for generating the most optimal Boolean expression. By varying the criteria and the values, you can apply this tool to select or prioritize many choices.

A Karnaugh Map is a 2x2 matrix. But it is also a participative tool that enables a group to prioritize or select items from a group of many, using a set of criteria. It is a simplification of a FMEA. In short, it enables you to make better and faster decisions.

Applications

Facilitators are often faced with helping a team make a decision or a selection to set priorities from an array of many options. For example, they must determine the risk factors of a new deployment, select key features on a new product, determine the importance of tasks or projects and much more.

The scientific community has been successfully dealing with this issue for some time. Over the years, engineers and scientists have developed tools and methods to help them cope with unanticipated failures. They have learned successful ways to anticipate what may go wrong in order to prevent it. From risk analyses to FMEA (Failure Model and Effect Analysis) to design of experiments, these methods, when well-understood and carefully applied, have proved to work.

Generally, when choosing we have two criteria to use — for example, "urgent" and "important." While some things can be very important, they are not necessarily urgent. Thus, the ability of a team to separate all tasks based on the set of criteria will result in a better deployment of limited resources.

Another example is risk selection. Take any project and you can ask your team to identify those things that are likely to go wrong, and the negative impact on the project if they do. These two criteria will help the team focus on those things that are likely to happen, and will cause serious damage to their project. Thus, rather than focusing on all the things that can go wrong, they can focus on the vital few.

Why 2x2 Matrixes

- The 2x2 matrix is attractive because it is graphical and simple and enables group participation.
- It engages everyone on the team, and allows for innovation and breakthrough.
- It can be as rigorous as you need it to be — i.e., the results can be made statistically defensible.
- It is very flexible, from prioritizing to selecting, choosing and more.

General Rules for Karnaugh Maps

- You should first be familiar with the concepts of Affinity Diagrams, or KJ (Kawakita Jiro) method.
- A session to develop a successful 2x2 map will take three to four hours.
- The challenge is to select the right criteria for each of the axes. The Y axis can represent one criterion such as urgency. You can assign values; i.e., a scale, such as 0 to 10.

Definition

Facilitator

Anyone who is leading a team to success; this includes project managers, team leaders, program managers, supervisors, leads, etc.

Pair-Wise

A process of comparing entities in pairs to determine preference.

The method is used in statistical analysis, specifically probability.

L.L. Thurstone first introduced this approach in 1927, using pair-wise comparisons for measurement.

For our purposes, a group or team must express a preference such as priority between two mutually distinct alternatives in order to meet an objective.

Karnaugh Maps: 2x2 Matrixes

- The X axis (for example, important) is your second criteria. Likewise, you can assign to it a value.
- Once you have a way to plot items based on the criteria (e.g., importance and urgency) place yellow stickies in the appropriate quadrant. For example, the items in the upper-right corner are more important and more urgent than the others.
- The color of the yellow stickies and their size can also represent additional criteria such as cost and difficulty of implementation. We recommend that you start with just two variables.
- We suggest that you limit your Karnaugh Maps to no more than three variables. Because Karnaugh Maps are a visual tool, it is best to limit the number of variables so as not to confuse the team.

Multiple Variables

Can you use more than three variables? Yes, you can. You can use a weighted-average method through the use of an Excel spreadsheet. You can have as many variables as you wish. But, it is important to show the results in only three variables. There is the tendency to complicate matters beyond what you need. Be careful! Let your objectives determine the number of variables. In our experience, most instances only require two or, at most, three.

The Scales

Generally, a scale of 1 to 10 on both axes is sufficient. It is important that the team agrees with the definition of the scale; i.e., decide what each incremental level means. Some teams prefer to use a 1 to 5 scale, which is also acceptable.

Remember, this approach is **QUALITATIVE**. A **QUANTITATIVE** scale is also acceptable, but we highly recommend that you choose these sparingly. The additional accuracy provided by a quantitative scale, in most cases, is overkill. Use quantitative scales when the topic at hand is critical and of extreme importance to your team or company.

A Better Way to Prioritize Affinity Diagrams

You have an alternative to help you prioritize your Affinity Diagram: You can choose the method outlined in the Affinity section (Prioritizing an Affinity Diagram) or you can choose the Karnaugh Map (KM) approach. Both are valuable and useful, although Karnaugh Maps provide more focus and perhaps a bit more accuracy. Karnaugh Maps also provide a more rational and organized approach than Affinity Diagrams. With Affinity Diagrams, you prioritize using a single variable, but Karnaugh Maps allow for two or more variables. For example, you could set the x and y axes to indicate "cost to develop" and "importance to the customer," and these two variables, called pair-wise, combine more than one criterion. You can even add a third variable; for example, the color of the stickies denotes "difficulty to develop." We suggest that you limit your Karnaugh Maps to no more than three variables. For example, the x axis is "cost to develop," the y axis is "importance to the customer" and the color of the Stickies is "difficulty to develop." In short, because Karnaugh Maps are a visual tool, it is best to limit the number of variable so as not to confuse the team. For examples of other combinations of variables, see examples of pair-wise criteria and their application later in this section.

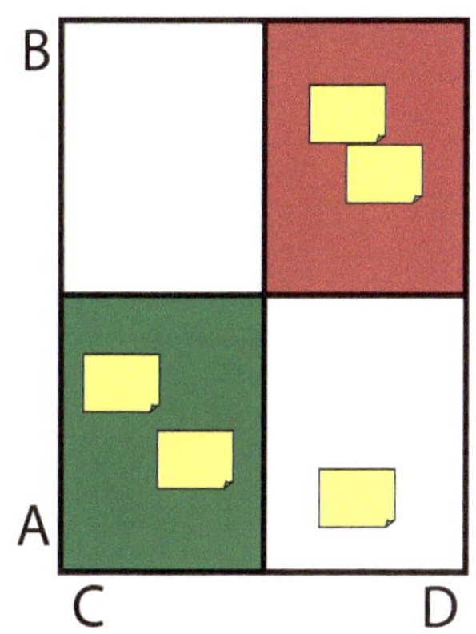

Typical configuration of a 2x2 matrix (Karnaugh Map). Notice the four quadrants.

Breadth of Uses for Karnaugh Maps

Once you learn to use them, you will find that they can be used in areas well beyond customer contacts. The famous 2x2 matrixes have been around for many years. They can be used to prioritize or make decisions. In short, any time a decision needs to be made, you can use Karnaugh Maps.

Karnaugh Maps: 2x2 Matrixes

Example: Karnaugh Map (2x2 Matrix)

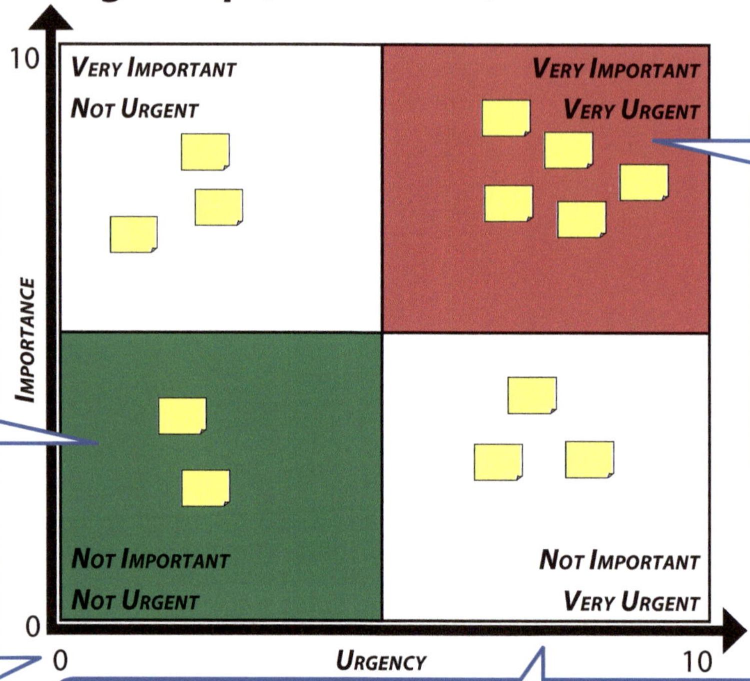

Items in the lower left quadrant are neither important nor urgent. So your team can abandon them or archive them for later use.

It is always best to set the lower left side as 0,0 points. This will create a more intuitive chart.

Yellow stickies in this quadrant represent the most important and the most urgent. Your team should focus on them by developing appropriate tactical or action plans.

The scale is determined by the intent of your diagram. It can be 0 to 10, 1 to 100 or Small, Medium or Large. For brainstorming sessions, the accuracy does not matter. For refinement purpose, you may need to make the scale more scientific.

Detailed Process: Prioritizing a Karnaugh Map

1

Set Objectives for Your Karnaugh Map

For example:

- Determine risks and rewards.
- Prioritize from a choice of many.
- Select options based on certain criteria.
- Choose a course of action, given certain criteria.

Important! It's a good idea to prepare a blank 2x2 matrix in advance. Tape two or more sheets of flip chart paper together and use markers to draw the matrix. A white board is an alternative, but upon completion you need to carefully gather the yellow stickies.

Important! After a brainstorming session, you may end up with as many as 300 yellow stickies. Scrubbing is the step to eliminate duplicates, redundancies and any other non-value-added yellow stickies.

Your goal is to end up with approximately 30 or fewer yellow stickies for the last step. On the other hand, if you end up with many more, and you feel they are important, then proceed.

Continued

Karnaugh Maps: 2x2 Matrixes

CONT'D

See Also Pair-Wise Criteria

2 Process the Information

Once all the yellow stickies have been collected, start scrubbing the information to ensure that every yellow sticky communicates a clear, distinct idea.

This step can take 30 to 45 minutes.

If you have too many yellow stickies, then you need to affinitize (group) them into main categories.

Important: The selection of the two criteria for the X and Y axes is critical. To help determine the criteria, spend sufficient time with the team discussing "why" or "how" the final information will be used.

3 Define the Criteria and Scale

Select the pair-wise criteria for the X and Y axes. For example:

- Important vs. Urgent
- Expense vs. Ease of Implementation
- Risk vs. Rewards
- Important to Customers vs. Expense to Develop

Set your scale. For example:

- Small, Medium and High
- 0 to 10 or 1 to 100
- A, B and C

Important: You can define a third criterion via the color or size of the stickies. Use this option sparingly. Furthermore, when first deploying the Karnaugh Map tool, we recommend you begin with just two criteria.

4 Plot the Information on the Matrix

Using your criteria, place each yellow sticky inside the 2x2 matrix. Do this in collaboration with your team.

You can use several methods of voting to assign value to each of the yellow stickies — for example, count hands, use dots, assign numerical values to later average. The more critical the topic, the more formal the vote should be.

Don't dispose of the plot, as you may need it for future reference.

5 Reach Your Conclusions

Placing all the yellow stickies does not complete the process!

The value of your 2x2 matrix is in the decisions that you make, the action-items that are generated and the follow through to act on these directives — simply put, the process is not over until something happens.

6 Develop Your Action Plan

Remember, it's not an action plan unless it clearly identifies "do what," "by whom" and "by when."

Document your action plan.

Follow up until all actions have been completed.

Karnaugh Maps: 2x2 Matrixes

Pair-Wise Comparison

Pair-wise comparison generally refers to any process of comparing entities in pairs to judge which of each pair is preferred or has a greater amount of some quantitative property. Pair-wise comparisons are relational — unlike Cartesian X and Y coordinates, where one is the function of the other. The method of pair-wise comparison is used in the scientific study of preferences, attitudes, voting systems, social choice, public choice and multi-agent AI systems. In statistical analysis, it's used to determine probability.

Prominent psychometrician L. L. Thurstone first introduced a scientific approach to using pair-wise comparisons for measurement in 1927, which he referred to as the Law of Comparative Judgment. Thurstone linked this approach to psychophysical theory developed by Ernst Heinrich Weber and Gustav Fechner. In psychology literature, it is often referred to as Paired Comparison.

Application to Product Development

In product development, your team can use pair-wise comparison to express a priority between two mutually distinct alternatives in order to meet an objective.

Many pair-wise criteria are possible. We offer a short list of examples to help you and your team develop criteria that fit your needs exactly.

1. **CHARACTERIZING RISK:** The criteria of "Likelihood of Happening (Probability)" and "Severity if it Does Happen (Impact)" — often expressed as a P-I Factor — maps risks that are most likely to occur and will have the most severity if they do. You can use the results to manage risk.
2. **SELECTION OF OPTIONS:** For example, when selecting projects or courses, use "Attractiveness" and "Cost" to determine best value.
3. **MAKING CHOICES:** When making trade-offs, try "Ease of Use" and "Cost of Implementation." "Achievability" and "Benefits" are simple criteria you could develop to have real meaning — for example "Achievable within the Fiscal Year" and "Financially Beneficial."
4. **DEFINING PRIORITY:** Use "Urgent" and "Important" to prioritize the allocation of resources.
5. **DOING A GAP ANALYSIS:** Use "Current Level versus Desired Level" and "Difficulty in Achieving."
6. **DEFINING COMPLEXITY:** Use "Complexity" and "Difficulty in Achieving."
7. **DOING RESOURCE ALLOCATION:** Compare "People" or "Resources" on the Y-axis with "Priority of Action" on the X-axis. The priority can be defined in strategic terms or some type of urgency scale.
8. **DEFINING THE COMPETITION:** Map the need to defend a segment, abandon it or create a new segment. One example: "Level of Competition" and "Urgency to Defend." Another example: "Level of Competition" and "Critical Market Segment."
9. **ENSURING FIT:** Use "Fit with Business Objectives" and "Cost of Implementation."

Karnaugh Maps: 2x2 Matrixes

Examples: Prioritizing Customer Requirements

Deliver the Value to the Customer Quickly

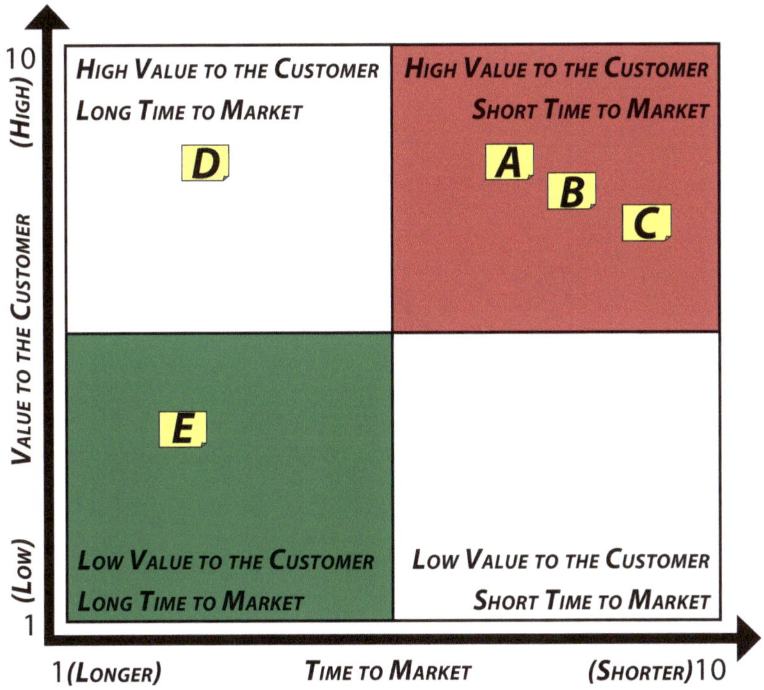

Here's an example of prioritizing based on pair-wise criteria: "Value to the Customer" and "Time to Market."

Items "A", "B" and" C" represent desirable candidates for high priority. Not only would these items deliver the highest value to the customer, but also they take the least amount of time to develop. In contrast, item "E" is not desirable at all.

Deliver the Value to the Customer in a New Way

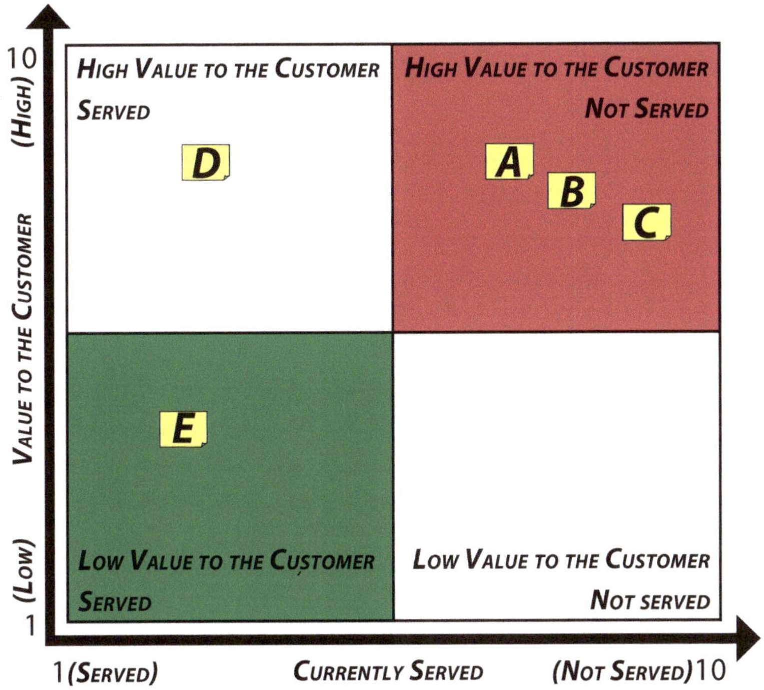

Here's an example of prioritizing based on pair-wise criteria: "Value to the Customer" and "Currently Served."

The variable "Currently Served" indicates how well the customer requirement is being addresses either by your product or by the products from your competitors.

Items "A", "B" and" C" represent desirable candidates for high priority. Not only would these items deliver the highest value to the customer, but also they are requirements that are not currently addressed. Again, item "E" is not desirable at all.

Awareness Curves

NEW TOPIC

Back

The Awareness Curve and Product Development

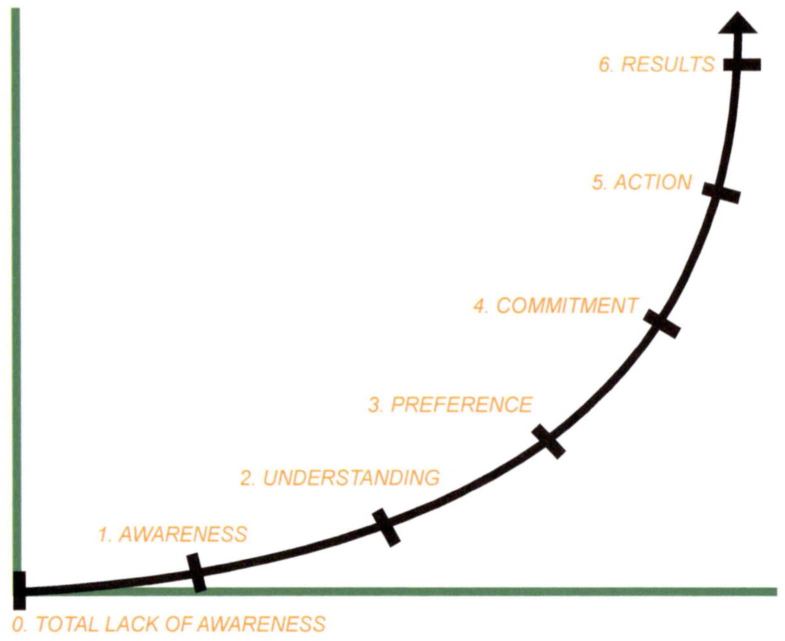

Figure: Awareness Curve

We go through a natural set of steps before we're willing to make a commitment to action. Social scientists and anthropologists have studied how we move from "not knowing" to "taking action." The Awareness Curve shows that there is a natural progression.

First we must be **AWARE (STEP 1)** of something. For example, if you don't know that your project has been approved for implementation, you are not able to support it.

From simple awareness, we seek additional information until we reach the next level of **UNDERSTANDING (STEP 2)**. Reaching this level varies from person to person, and some need considerable additional information to move from awareness to understanding. This fact should influence your communication strategy, specifically that you need to provide the right amount of information. For example, now that your project has been approved, you need to communicate to your team how they are affected.

Once understanding has been reached, we develop **PREFERENCE (STEP 3)**, which is the positive or negative image of the information. For example, you need to work at creating the positive preference that your project has been approved for implementation; it does not happen by itself. Talk to people, champion your project and start sending information.

Out of preference comes **COMMITMENT (STEP 4)**, and it is not until one reaches commitment that he or she is ready to take **ACTION (STEP 5)**. Naturally, action leads to **RESULTS (STEP 6)**, which is what you, as the project manager, are trying to achieve.

You are ready to deliver results. But consider that you are at the action level, while others on your team may still be at the total lack of awareness level. This means that they don't even know there is a project, let alone that it has been approved for implementation. These people will be reluctant to take any action, as they must first satisfy their needs for awareness, understanding, preference and commitment. You need to appreciate this gap!

Do you want to reduce chaos or complexity? Develop a plan to communicate effectively in order to bring all team members to the level of commitment. When new members come on board, pay them the same courtesy.

For comments, questions or additional copies please e-mail the author: flexpm@rapidinnovation.com.

The Ladder of Abstraction

The concept of the Ladder of Abstraction has its origin in the study of semantics. It was first postulated by Alfred Korzybski, but its popularity is attributed to S. I. Hayakawa in his work "Language Thought and Action" (1990, fourth edition). It is a very powerful concept that permeates all levels of society, yet few people truly apply it. Naturally, it is critical to customer interaction.

We have all read claims in advertisements like this: "it cleans 30 percent better!" Better than what? The statement is senseless because it does not provide context or reference. Many of our communication problems stem from people using a language that contains details that are different from the level of detail of the person doing the listening.

For example, the CEO of a company speaks in terms of dollars, while a technician speaks in terms of nodes in a circuit. You can visualize that these two individuals would have difficulty communicating. The CEO is worried about the price of the stock, and the technician is worried about cross-talk in her circuit design. These two people then are said to be at two different levels of abstraction. The wider the gap, the more difficult communication becomes.

Critical Principles of the Ladder of Abstraction

- Our reality is described by levels of abstraction.
- In a Ladder of Abstraction, all levels are "true" at the same time.
- The ladder has a "point of convergence," where the two parties find common ground and can communicate.
- Often people argue only because they are at two different levels in a Ladder of Abstraction.

Process

Learn how to build ladders of abstraction. They are simple but provide considerable information. They also allow you to rise to whatever level your customer chooses.

When interviewing customers, focus on the problems they are trying to solve. The critical strategy is to focus on why your customers do what they do.

Avoid arguing with your teammates when the only difference is the level of the ladder.

The Ladder of Abstraction

Examples: The Ladder of Abstraction

Example 1

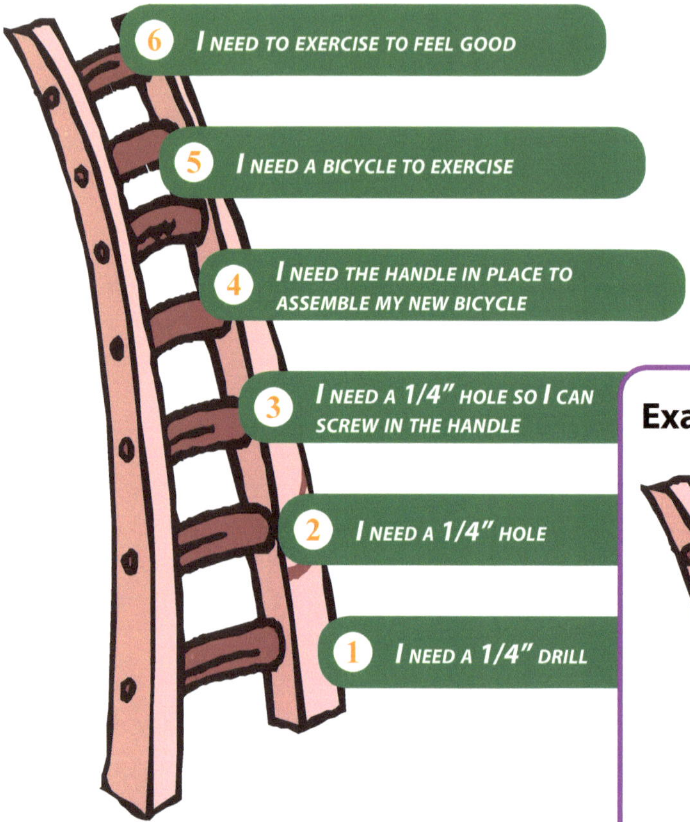

If you start reading from the bottom of the ladder toward the top, you will notice that, as you move up, the concepts become more strategic (abstract). Thus, what appears to be a mundane need for a 1/4" drill is actually driven by a very strategic need for well-being.

The rule states that all the items listed above are true at the same time; also, that things become more abstract (strategic) as you move up the ladder.

Example 2

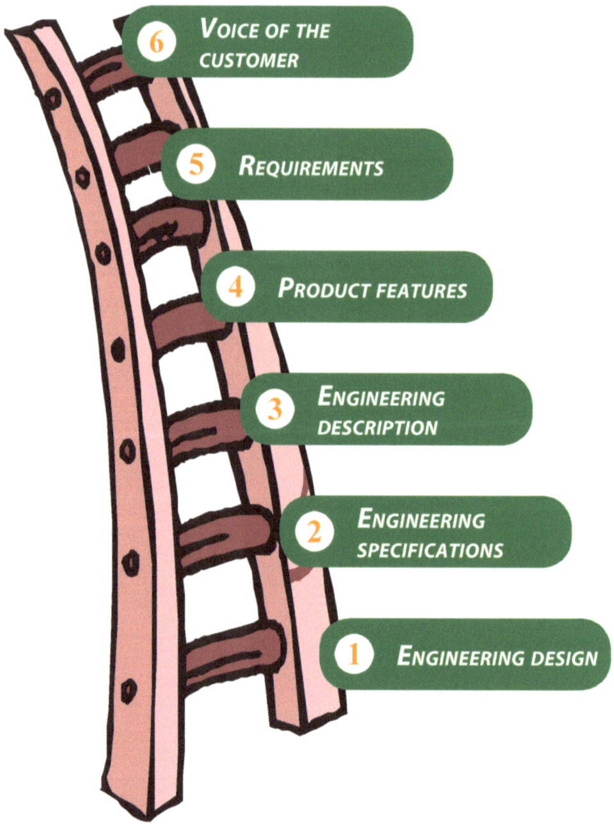

In this example, start reading from the top. Customers express their needs abstractly — for example, "make it easy to use." Going from Voice of the Customer to engineering design is a journey through a Ladder of Abstraction. You will need to turn the Voice of the Customer into a set of requirements, which then become features, and so on.

NEW TOPIC

The 5 Whys

Help Customers Reflect Their True Needs

This is a simple methodology to obtain valuable information from customers. As the name implies, in the literal sense, it means asking the customers "why" five times in order to get to the root message — i.e., the most fundamental message concerning the needs of the customer. This methodology originated in Japan and is typically associated with the process of identifying why failures occur. It is part of Root Cause Analysis.

In the context of customer interviews, the 5 Whys serves the purpose of getting past the stock answer. It allows you to peel away the layers of information until you get the valuable information underlying the needs of the customer. The 5 Whys is an excellent tool to get your customers to tell you their unarticulated or unstated needs. The process of asking why causes the customers to reflect on their actual needs.

General Rules for the 5 Whys

1. You can use as many whys as needed. While the name of the methodology is "5 Whys," it simply means that you can continue to ask why until you get useful information. In some cases, three whys will suffice, yet in other cases as many as seven will be needed. Use your judgment, and stop asking when you feel you have obtained the root message or the valuable information you are seeking.
2. Avoid the toddler's curse, when a 3-year-old child drives you crazy asking "why." Instead, let the customer talk, then ask why. You can also use other ways to ask — for example: "Tell me more" and "Why is that?"
3. Help the customer by restating her previous answers or by repeating the questions. For example: "You said that this product is difficult to use, and that this is because the menu system is not intuitive; now, can you describe why the menu system is not intuitive?
4. The intent of the 5 Whys is to get to a level where you get actionable input and not generalities. Actionable input is customer information that you can actually use in developing requirements and features.
5. The 5 Whys is not intended to be used as an interrogation tool. That is, one does not actually ask "why" five times. We encourage you to use your own style to apply the concept.
6. Your role is to enable your customer to talk and express his "root message."

The 5 Whys

Examples: The 5 Whys

Example 1

INTERVIEWER: How do you feel about this software product?
CUSTOMER: It is very difficult to use.

INTERVIEWER: Tell me more,
CUSTOMER: The menu system is not intuitive.

2nd Why
INTERVIEWER: Tell me why it's not intuitive.
CUSTOMER: I get lost all the time.

3rd Why
INTERVIEWER: Why do you think you get lost using the menu?
CUSTOMER: Well, there are too many layers and no way to get back.

A
ROOT MESSAGE: The menu system has too many layers and no way to get back, which makes customers feel that the product is too difficult to use.

Note the root message — i.e., actionable input from the customer.

Example 2

INTERVIEWER: Why did the fire start?
CUSTOMER: The problem was a short in the main transformer.

INTERVIEWER: Why did the transformer short?
CUSTOMER: Water got inside the mechanical package of the transformer.

2nd Why
INTERVIEWER: Why did water get inside?
CUSTOMER: The gaskets froze, became brittle and cracks developed, allowing water to seep.

3rd Why
INTERVIEWER: Why did the gaskets crack?
CUSTOMER: They are not specified for cold weather.

A
ROOT MESSAGE: The gaskets specified in the design document did not meet the environmental conditions where the transformer was placed.

The following examples illustrate the concept of the 5 Whys applied to the important task of interviewing customers

Examples: The 5 Whys

Example 1: The Main Pump Failed

Q

Q: Why did the main pump fail?

A: The coupling from the main intake to the pump broke and created a shutdown of the line.

1st Why

Q: Why did the coupling from the main intake fail?

A: The "O" ring failed under pressure.

2nd Why

Q: Why did the "O" ring fail under pressure?

A: The night temperature fell to -20 degrees and the "O" ring was not specified for low temperatures.

3rd Why

Q: Why was the "O" ring not specified for low temperatures?

A: No one expected these pumps to be sold in Alaska.

A

ROOT CAUSE: Insufficient environmental specs to enable safe operation of the pump under arctic conditions.

CORRECTIVE ACTION: Review all environmental specs and redesign for arctic conditions.

Example 2: The Machine Operator Received an Electric Shock when Using the Pick and Place Machine

Q

Q: Why did the machine operator received an electric shock when using the Pick and Place machine?

A: The operator touched the high voltage line.

1st Why

Q: Why did the operator touch the high voltage line resulting in shock?

A: The safety cover to the high voltage lines was open and the operator reached inside.

2nd Why

Q: Why did the operator reached inside the cover?

A: The operator thought that there were adjustment controls inside the cover.

3rd Why

Q: Why did the operator think that there were adjustments inside the box?

A: The box had no label and was opened.

A

ROOT CAUSE: Lack of proper safety marking and also a safety box that has no lock nor "Poka Yoke."

CORRECTIVE ACTION: Place warning labels prominently on and about the safety box. Add a lock that only the maintenance technicians can use. Add an automatic disconnect when the box is opened.

The following examples illustrate the concept of the 5 Whys. In both cases there are only three follow-up questions. In some cases, it could take seven or more follow-up questions. The 5 Whys concept can be used in many areas of inquiry.

Conflict and Conflict Management

NEW TOPIC

Conflict and Conflict Management

Recognizing Conflict

Conflict in product development is inevitable; we define conflict as a disagreement, or a controversy. For example conflict about the development schedule, where marketing selects a date and engineering opposes it.

Three Sources of Conflict

1. Perception: A misunderstanding, the way people perceive reality is colored by their own biases, such as culture, religion, aptitudes, etc.
2. Emotion: emotional attachment to a particular direction or path, which collides with some else's passion. For example, a "pet feature" or a dislike of a particular type of customer.
3. Communication: Lack of clarity or inability to communicate. For example, when writing to some one whose native language is not English.

While we should all strive to avoid conflict, total harmony may not be achievable in the fast-paced world of product development. Consequently, we must be equipped with the ability to deal and resolve conflict at the earliest possible time.

How People React to Conflict

- Confrontation: An open conflict of opposing ideas. While confrontation is not a bad thing, it helps address problems that must be resolved, it is none the less disconcerting to some people as it comes across as belligerent or ready to fight.
- Avoidance: Some people chose not to recognize or admit that there is problem which must be solved. They try to deal with it by withdrawing or pretend that the problem is not there.
- Negotiation: Some people are eager to resolve a conflict through negotiation, hoping to obtain some gain as a result.

Dealing with Conflict

- Stay objective, that is manage by fact and data. When people use anecdotal information the conflict will most likely escalate to a serious breakdown, maintain your objectivity. Many times, conflict revolves around a "symptom"; for example, "there are too many field failures", may create a conflict by blaming engineering for a poor design, while it may be a defective part shipped by mistake by one of your vendor, use "root cause analysis" methods to avoid this type of problems.
- Communicate with clarity: While this is an obvious way to manage conflict, we tend to underestimate its important. Many conflicts are due to poor communication; for example a poorly written email is misunderstood by the recipients. Improve your communication skills, understand your audience and take into consideration cultural differences.
- Validate: Verify what the conflict is about, it may not be a conflict at all. Many times we argue about things we already agree upon. Use verification techniques, ask people to state the problem with clarity, then repeat it to ensure that what they said and what you heard are the same things.
- Separate the people from the problem: You are better able to stay objective if you avoid personal attacks or the laying of blame on a particular individual. At the time of conflict, the only issue is to resolve the problem regardless of whose fault it may have been.
- Develop a "win-win": Resolve the conflict in a way, if possible, that there is mutual gain.

Above all, resolve the conflict to enhance the objectives of your development program, remember that the project should always win!

Conflict and Conflict Management

Negotiations

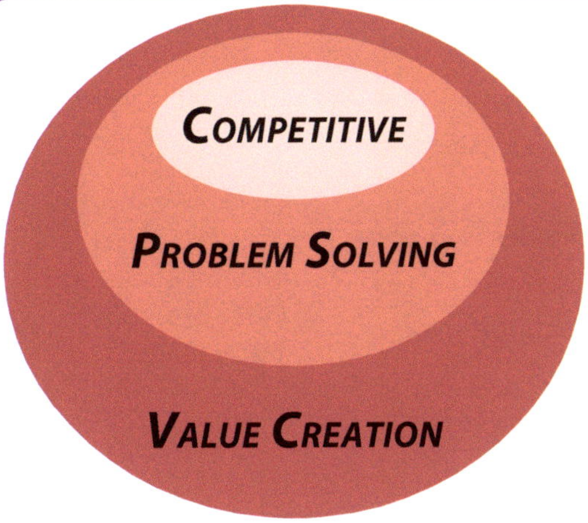

This is a mental model of the three types of negotiating styles. Negotiating is all about making a decision to ensure positive progress in your product development project. We favor the value creation approach, as it reflects a collaborative environment where the intent is to quickly decide to create a win-win, but remember your project always wins.

The **COMPETITIVE** type is not desirable, as it aims to have one winner and one loser. It complicates the negotiation, as the outcome is blurred by the desire to win rather than find a satisfactory solution.

PROBLEM SOLVING is a better choice, as it strives to create a collaborative environment where everyone is focused on solving a problem regardless of who wins and who loses.

The best strategy is **VALUE CREATION**, where negotiators are seeking to innovate and find better solutions; everybody's energy is intensely centered on discovering many alternatives, improvement and innovation.

Common Types of Negotiations

	Competitive	Problem Solving	Value Creation
Description	• Seeks to maximize your own position. • Seeks to control the negotiation process. • The ends justifies the means.	• Considers needs of the other side as relevant. • Tries to discover and share joint gains. • Collaboration is the preferred approach.	• Sees negotiation as attempting to create new shared value for all participants. • Focus is on needs and interests of the partners.
Application	• Simple problems. • When only a single resource involved.	• When it's possible to articulate common perspectives among equals on joint problems. • When shared resources can address a solution.	• Best for solving tough problems where the interest of the parties differ. • Problems requiring innovation, and breakthrough solutions.
Advantages	• Low context. • Quick.	• Practical to use. • Most common form of negotiation in product development. • Allows for collaboration. • Creates trust for future negotiation.	• When there is a common purpose, there is no better way to negotiate • Joint accountability. • The process creates "buy-in."
Disadvantages	• Limited solution space. • Can create distrust.	• Takes more time than competitive negotiation. • Can be abused by those higher up in the organization.	• Overly complicated for simple problems. • Neutral, no clear winner or losers.

Conflict and Conflict Management

Keys to Facilitating Negotiations

- Become aware that you are about to negotiate.
- Prepare — it's absolutely critical.
- The goal is quality business decisions.
- Recognize the tactics of the negotiators, and set the right context. Match their style of negotiation.
- Try to shape the process and context agreements to enable all sides to choose tactics appropriate to the issues at hand.
- When you are at loggerheads, seek facts from both sides that support their assertions.
- Take a coffee break if you reach an impasse, allow time for alternatives to form. Incidentally, the break can be a few minutes or even a few days.

Important

A team should have a common purpose, one which is agreed to at the start of the project. They need to refer to their common purpose at the start of the negotiations, and continue to use it to bring alignment.

Tips and Hints

- Separate the people from the problem.
- People problems that do occur: perception, emotion and communication.
- Invent options for mutual gain.
- Insist on using objective criteria to evaluate the solution or agreement.
- Focus on interests and not positions.
- Humans like to be heard. Give your interests and reasoning first, but save your conclusions and proposals until later.
- Look forward not back. Focus on the future; don't cry over spilled milk. There is no use reminiscing for hours-on-end!
- Attack the problem without blame. Blame serves no purpose; it only aggravates communications and stirs negative emotions.
- Always strive for a win-win. It improves relationships, creates alignment and avoids confrontation.
- Discuss each others' perceptions. Encourage the exchange of perceptions and understanding of the situation and the topic of the negotiation.
- Foster understanding. Negotiators tend not to talk to be understood. Encourage everyone to seek understanding first.
- Questions will carry the day. Encourage everyone to ask questions because it is the most powerful way to reach understanding.
- Write things down. If at all possible, write on the flip chart or whiteboard. Write down the questions, the topic of negotiation, perceptions and positions. Written information increases understanding

Yellow Sticky Protocol

Effective Group Participation

Benefits

The ability to ensure that everyone in a group participates is greatly enhanced through the use of Post-It Notes® (trademark of 3M), known in the vernacular as "yellow stickies." Another benefit of yellow stickies is that they provide a self-documenting process; rather than writing things on a flip chart, which is time-consuming, yellow stickies significantly accelerate the documentation. Finally, yellow stickies are portable; that is, one can move them, hand them over, rewrite them, etc.

Some individuals, particularly senior executives, are not fond of yellow stickies. This is due to a misunderstanding as to who uses them. But this is a serious tool that will produce better results, and more quickly. These people need to be reassured that, barring a technological breakthrough, yellow stickies are still a fine tool that everyone, including executives, can use.

The process favored in this guidebook relies heavily on the proper use of yellow stickies.

Essential

"The process favored in this guidebook relies heavily on the proper use of yellow stickies."

Establish the Protocol

To derive the benefits of using yellow stickies, the facilitator must establish a protocol — an articulated set of guidelines as to the way the information will be presented in each yellow sticky.

The protocol is a way to create a safe environment. For example, use the protocol to assuage members who may have reservations about their poor spelling, grammar or handwriting.

The protocol is a way to choreograph the process so it yields maximum benefits.

For the sake of productivity, let alone sanity, the protocol must be enforced. All team members must participate and follow the protocol. Every yellow sticky must conform to a standard to be useful.

There is flexibility, so you need not do it exactly as in the corresponding detailed process. The overall goal is to be able to track and process a large amount of information.

Your team must agree on the way the yellow stickies will be structured. Once you have reviewed and highlighted all of your notes, you proceed to transfer the information to the yellow stickies. You need a protocol as to the type of information and its location. This is critical to provide continuity, traceability and consistency

Plan Ahead

Schedule a three to five hour meeting for a team of up to 15 people in a large conference room. Prepare topics and objectives for the meeting as well as the introduction. Gather materials:

- Post-It Notes® (3M brand, model #655 3x5).
- Sharpie® (Sanford brand, fine point, black markers) for all attendees.
- Flip chart paper and easels — two are best.

Level 3 Detailed Process: Yellow Sticky Protocol

1

Prepare your Materials

- 3x5, 3M-brand, Post-It Notes® (#655) — One pad per participant. Consider that several of the participants will have a sizeable number of yellow stickies — in some cases as many as several hundred.
- Sanford-brand Sharpie® markers, fine point, black.
- Flip charts or a place to hang the yellow stickies once filled.

Important: You also will need the participation of everyone who made customer visits.

2

Make the Announcement in Front of the Group

Face the audience, make eye contact and say...

"We are going to use yellow stickies in this meeting.

There are three key reasons why this is important:

- It is a self-documenting process, so it saves time and allows for documentation after the meeting.
- It allows everyone to participate.
- We can move the yellow stickies around depending on our needs — no need to rewrite, just move them.

Important: Caution: Some individuals, particularly senior executives, are not fond of yellow stickies. This is due to a misunderstanding as to who uses them. But this is a serious tool that will produce better results, and more quickly. These people need to be reassured that yellow stickies are a fine tool that everyone can use.

3

Explain the Protocol to the Group

"In order to take full advantage of the tool, we must follow a simple set of rules..."

For example:

1. Don't worry about spelling or grammar; no one will check it. This goes for everyone in this room.
2. Don't write on the sticky side of the note.
3. Write landscape (horizontally) oriented.
4. Use your Sharpie® marker. This will force you to write larger than usual and more legibly, and it will enable others to read notes from a distance.
5. If you make a mistake or change your mind, simply toss it and start a new one.
6. Write only one item per yellow sticky. Only one question, one idea, one quote.

Important: Encourage everyone to write legibly. Ask them to print. Many of us do not have the best of penmanship. Explain to the participants that they are writing for others to read! It will save time and reduce confusion.

4

Start the Process

See Also: Affinity Diagrams

Example: Yellow Sticky Protocol

The topic of conversation from your discussion guide.

This must match the topics contained in the discussion guide that was created for this set of customer interviews.

An agreed-to indication of the importance, relevance or some other priority. In this example, "H" indicates "High" passion or importance of the statement from the source, whereas "M" would indicate "Medium" and "L" would indicate "Low."

Your team can develop similar priorities, depending on the objectives of your visits. Having a priority set will save considerable time during the later stages of this process. Keep it simple; otherwise, your team members will be confused and misinterpret the code.

This area is reserved for the message from the interviewee — the voice of the customer.

It can be a quotation, a fact, a finding, information or any sentence that is relevant. This is extracted from the notes you took during the interview — you must be able to trace it to your notes.

The Protocol

Topic H Customer's title

"Actual quote from the customer, extracted from the interview notes."

Your name/initials Page, Interview #

The upper right corner is for the title or function of the person interviewed.

You need to develop codes to be used here to help you understand the context, particularly if your team interviewed customers from different functional groups.

The customer's name is not relevant.

Quote marks (" ") are reserved for actual customer quotes! This is a critical rule.

Quotes are always verbatim messages from the customer. Do not use quotes if the customer did not actually say it.

A simple code will enable you to trace the information to your notes, and thus to an actual interview.

It is difficult to remember who said what. During the processing and various meetings, you will want to refer to your notes to clarify the context of the statement.

This is simply a tracking mechanism. You should be able to track back to the actual source of the information.

Example

Ease of use H Design engineer

"More than three layers of menus make products difficult to use."

Fred T Page 1, 2nd interview

Yellow stickies are landscape (horizontally) oriented)

This example shows a super set of the Yellow Sticky Protocol. You need not use all of the items, just those that will enable you to track and code the critical information. When you have several hundred yellow stickies, you will rely heavily on the protocol to make sense of it all.

Affinity Diagrams

NEW TOPIC

Organize Information and Reach Consensus

Introduction

Affinity Diagrams are a very practical, easy to use method that can be applied in many areas once you learn how to facilitate the process — you lead your team through the process, and your team members look to you for instructions and guidance to reach the results.

Affinity Diagrams can be used in many areas of team work:

- Prioritize topics from a list of many
- Collaborate to solve a problem
- Develop a strategy
- Define group options and make choices
- Identify alternative paths or options
- Brainstorm discuss a set of issue

In short, Affinity Diagrams allow you to tap the power of a team by enabling everyone to participate and ensuring that everyone has the opportunity to innovate and create.

Definition
Affinity
A natural liking for or inclination toward somebody or something or a feeling of identification with somebody or something.

Definition
Affinity Diagrams
A tool to organize information and reach consensus."

Essential
"Affinity Diagrams are essential to the process favored by this guidebook."

Historical Background

Affinity Diagrams belong to a set of tools called the KJ method, which is named after Kawakita Jiro, the Japanese anthropologist who first introduced the approach in 1967. The approach has its roots in the values of Zen Buddhism.

Mr. Kawakita developed the method as a means of organizing diverse observations and qualitative information into useful, documented facts. He observed that humans are not linear thinkers; that is, ideas flow in a non-linear fashion. Further, humans tend to jump around when brainstorming. Thus, trying to structure brainstorming sub-optimizes the creative flow.

Why Affinity Diagrams

- They are simple and non-threatening. As the saying goes, "They're not rocket science." This creates an inviting environment for full participation.
- They are inclusive. Properly used, an Affinity Diagram allows everyone on a team to fully participate, regardless of stature or pecking order.
- They are a fine way to ensure consensus in a group, as well as to build the sense of team.
- They are a practical way of organizing seemingly unrelated facts.
- They are very conducive to the creation of clarity and communication within the team.

General Recommendations for Affinity Diagramming

1. Planning your session is essential to success. Please do not treat an Affinity Diagram process as a casual affair. Consult the process map in this document for additional information.
2. Select a room large enough to allow the team to move around. You may have to move furniture to create a clean space to work.

Affinity Diagrams

3. Affinity Diagrams work best when done by a group of about 10 people. Larger groups tend to fracture, meaning that some tend to marginalize themselves and not participate. Thus, groups larger than 10 require considerable facilitation and constant focus to ensure productivity.
4. Silence or talking? The general rules indicate the whole process should be done in silence to allow creativity to flow and avoid any one individual influencing the group. We believe that, in certain circumstances, talking is desirable. When you have a small group, say about three to four people, or when everyone is a peer and collaborative by nature, these are examples of when talking might help. Use your judgment and decide based on the dynamics of your group.
5. Use the traditional rules of brainstorming — for example, do not allow criticism of ideas, encourage participation by all team members and generate the largest possible number of ideas, including duplicates and repetition.
6. Remember the follow-up. Many fine sessions end up delivering little or no value due to the lack of action after the session. Ensure before you start that you are committed to following up well after the session is completed. Every session must have an action plan ("road map") to ensure that you produce tangible results.
7. It is best to use flip chart paper. We also recommend other types of large-format paper such as butcher paper, newsprint rolls and _. Do not hesitate to create a large space with paper. Remember that your team will be working in this area — six to nine horizontal linear feet of paper is best.
8. Avoid cryptic descriptions, such as one-word statements like "energy" or "easy." Encourage team members to use the proactive or active tense — for example, "There is not enough energy" or "Make it easy to use." Therefore, statements should have a noun and a verb to avoid ambiguity and enhance clarity.
9. The number of categories or columns should be kept to about five; more than that confuses the teams.
10. Create the feeding frenzy. It is fun and productive to encourage the team to generate as many ideas as possible. Set challenges — for example, ask everyone to generate five ideas. Do many rounds, create contests and give a prize to the person who generates the most ideas.

High-Level Process

1. **Start**
   Deliver instructions.
2. **Brainstorm**
   Participants create yellow stickies.
3. **Post**
   Place yellow stickies on the wall.
4. **Organize**
   Participants organize yellow stickies into categories.
5. **Prioritize**
   Participants prioritize categories.
6. **Develop Action**
   Assign responsibilities and develop a timeline.

Important This is a high-level view of the process for an Affinity Diagram. It will help you navigate, without the complexity of all the details; i.e., the Detailed Process Map.

Affinity Diagrams

Detailed Process: Affinity Diagrams

1

INVITE THE PARTICIPANTS

Who should attend?

- All stakeholders in the process or project.
- Customers, if appropriate.
- Try to keep the group to 10 participants or fewer.

2

PREPARE

- A large conference room
- Post-It Notes® (3M brand, model #655 3x5)
- Sharpie® (Sanford brand, fine point, black markers) for all attendees
- Flip chart paper and easels — two are best

See Also: Yellow Sticky Protocol

3

SET UP THE TOPIC

- State the topic in the form of a question. Ensure that everyone understands it. You may want to discuss it with the group.
- Reassure everyone that there are no wrong answers.
- Reassure everyone that grammar and spelling are not important.
- Explain how to use the yellow stickies — one per subject, use a Sharpie® pen, landscape (horizontally) oriented.
- No criticism.

4

START THE PROCESS

- Give the question or topic, then ask everyone to write at least three yellow stickies and wait.
- Ask each person to read his yellow stickies as loudly as possible.
- Tell everyone to listen intently as others read their yellow stickies.
- Encourage participants to write additional yellow stickies as others present theirs.
- Collect all yellow stickies and place them on the flip chart or board at random.
- Continue until the group runs out of ideas or repetition sets in.

5

SCRUBBING THE AFFINITY DIAGRAM

Clean up your Affinity Diagram once all the yellow stickies are written and placed on the board.

- Eliminate yellow stickies that are obviously irrelevant or trivial.
- Eliminate obvious duplications.
- Rewrite yellow stickies that are difficult to read.
- Rewrite yellow stickies that are not clear.

Important: For any and all changes you make when scrubbing the Affinity Diagram, make them with the agreement of the authors.

CONTINUED

Affinity Diagrams

Cont'd

6 Start the Affinity Process

Rally the group to move up around the flip charts with the yellow stickies.

Ask participants to observe how the various yellow stickies cluster or group into subjects.

Ask the participants to start moving the yellow stickies, using their own criteria to cluster them, regardless of how others feel about it.

Do this in silence! No one can talk during this step!

Stop when there is little or no movement of yellow stickies.

7 Pick 'n Choose

Review each column or cluster that was created in the previous step, with participation by the group.

Ask the group to explain why the cluster or column exists. "Why are these yellow stickies together?"

Move the yellow stickies that do not belong in the column to a different one or a separate place on the flip chart.

Eliminate duplicates by choosing one that represents the topic and placing the others underneath.

For each column, assign a header that captures the topic. Write the header on a sticky notes, preferably a color other than yellow.

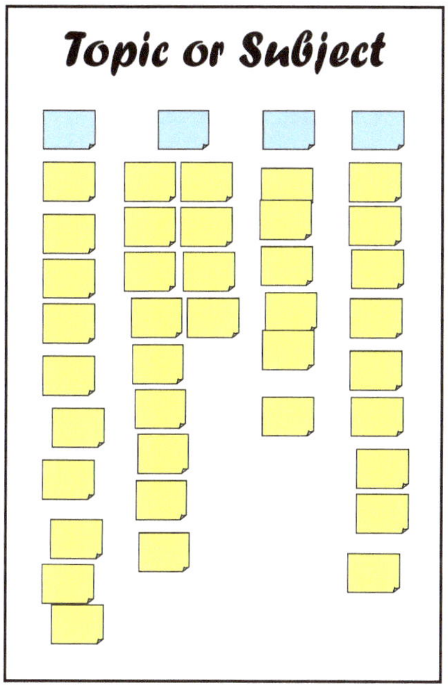

Yellow Sticky Approach

8 Prioritize

Once the notes are prioritized, you can begin to assign responsibilities and action items as appropriate.

Also, develop a timeline. This is a critical step!

Save the flip charts for archival purposes or to be transcribed if necessary.

Debrief the group on the effectiveness of the tool and ways to improve the process.

See Also: Prioritizing Affinity Diagrams

See Also: Karnaugh Maps

Prioritizing Affinity Diagrams

NEW TOPIC

The Process to Prioritize

Most often, Affinity Diagrams are used to prioritize a set of ideas, choices, things, etc. The criteria for prioritizing are determined by the initial objective that you set for the Affinity Diagram — for example, you might want to do an Affinity Diagram to assign responsibilities or action items. You might also want to pick a few from a selection of many. In short, use the same objective you set for the Affinity Diagram to select the method you will use to prioritize.

Group Consensus

Prioritizing is all about making decisions; keep this in mind as you work with your team. You want to reach consensus. 100 percent agreement is desirable, but not essential. Most of your prioritizing efforts will not have 100 percent consensus. That being said, prioritizing an Affinity Diagram is a team effort.

Select The Criteria

The key to prioritizing is the selection of the criteria used to make the decisions. Take your time, discuss with your team, and above all ensure absolute clarity. Criteria should have two or three elements; in a team environment and without computers it's very difficult to keep track of more than three criteria.

This process should be driven by the strategic importance of the project. For small, simple projects the criteria might be somewhat subjective. If the project is critical, then the criteria must be defensible — objective, verifiable and in full alignment with the strategic goals of your project. This is a qualitative exercise, so avoid criteria that are too complex.

Common Methods

1. ABOVE OR BELOW THE LINE

Draw a horizontal line at the midway, and ask the team to decide for each yellow sticky if it should be above or below the line. This is the quickest approach and one that most people find intuitive. You can prioritize the yellow stickies or the headers depending on your objectives.

2. 1 ROUND OF DOTS

This is a popular method because it's simple and familiar. Give everybody five to 10 self-stick colored dots to place on the yellow stickies based on the predetermined criteria. They can place their dots anywhere, as many dots as they wish per item. When done, simply count the dots to establish priority.

3. 2 ROUNDS OF DOTS

A variation on this theme is to do two rounds of dot placing with five dots for both rounds:

Round 1: Place five dots anywhere, with no guidelines.

Round 2: Place five dots, but only on two yellow stickies.

4. 3 ROUNDS OF DOTS

A more sophisticated method involves giving each team member 30 dots to be placed during three rounds:

Round 1: Place 10 dots anywhere, with no guidelines.

Round 2: Place 10 dots, but no more than two dots on any one yellow sticky.

Round 3: Place the remaining 10 dots, but only on two yellow stickies (five and five).

5. YOUR INVENTION

Invent your own system for prioritizing. As long as it's rational and leads to clear decisions, you'll be fine.

Level 3 Detailed Process: Prioritizing Affinity Diagrams

1

Select the Criteria

The set of criteria is generally two to three elements that will determine priority. The criteria will drive the decision process.

Take your time, discuss with your team, and above all ensure absolute clarity.

The simplicity or complexity of your criteria depends on the strategic importance of the project. As a rule, keep it simple.

Once agreement has been reached, write the criteria on a flip chart to help the team stay consistent.

Important! This process should be driven by the strategic importance of the project. For small, simple projects the criteria might be somewhat subjective.

If the project is critical, then the criteria must be defensible — objective, verifiable and in full alignment with the strategic goals of your project.

2

Start the Process to Prioritize

Q. Do we prioritize the headers, or each yellow sticky?

A. It depends on your objectives:

The preferred method is to prioritize the individual yellow stickies. Based on which categories the prioritized yellow stickies are in, you can infer the priority for the categories.

For example, if you use one of the Dots methods, count the number of dots in each category, and write it on the header. This will create an overall priority ranking.

Another approach is to ask the participants to place the dots on the actual headers — not the yellow stickies. This is faster, but it may miss an important message: Which yellow sticky is most important?

Use your judgment as to which voting method will serve you best.

Important! Prioritizing is a team effort. 100 percent agreement is desirable, but neither essential nor expected. The goal is consensus.

3

Prioritize

Choose from one of these common methods or invent your own:

- **ABOVE OR BELOW THE LINE:** Draw a horizontal line, and ask the team to decide for each yellow sticky if it should be above or below the line.
- **1 ROUND OF DOTS:** Give everybody five to 10 self-stick colored dots to place on the yellow stickies based on the criteria. They can place their dots anywhere, as many dots as they wish per item. When done, simply count the dots to establish priority.
- **2 ROUNDS OF DOTS:** A variation on this theme is to do two rounds of dot placing with five dots for both rounds:
  - Round 1: Place five dots anywhere, with no guidelines.
  - Round 2: Place five dots, but only on two yellow stickies.
- **3 ROUNDS OF DOTS:** A more sophisticated method involves giving each team member 30 dots to be placed during three rounds:
  - Round 1: Place 10 dots anywhere, with no guidelines.
  - Round 2: Place 10 dots, but no more than two dots on any one yellow sticky.
  - Round 3: Place the remaining 10 dots, but only on two yellow stickies (five and five)

Prioritizing Affinity Diagrams

Example: Affinity Diagram

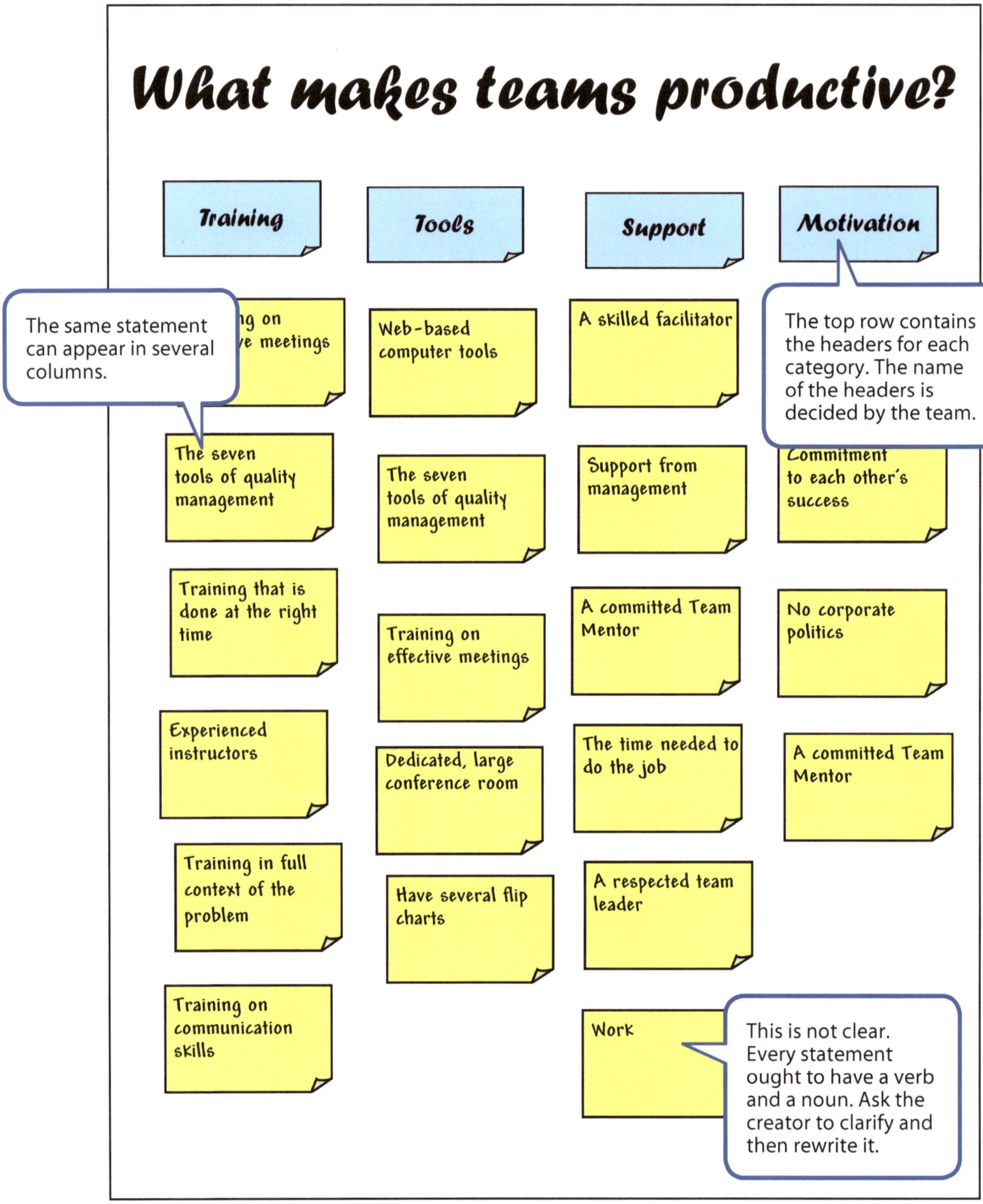

The Fishbone Diagram

Diagramming Cause and Effect

Generally, when something goes wrong — for example, a machine stops working — we tend to jump to conclusions based on our experiences and immediately infer the reason for the failure. Worse yet, we go all the way to a solution. There is a risk in this behavior. What if the reason for the breakdown is not what we thought? What if our solution is the wrong one?

The Fishbone Diagram, also known as the Ishikawa Diagram, is a fine tool for providing the discipline to go from "symptom" to actual "root cause" of the problem, regardless of the nature of the problem.

Teams are frequently confronted with solving problems; i.e., addressing a snag or an obstacle. Furthermore, the definition of the problem tends to be, in reality, a symptom of something else. For example, "the press does not work" or "the car does not start."

It is very tempting to try to solve the symptom of the problem without trying to analyze the real cause. In the example above, "the press does not work" may lead you to try several fixes, only to find out that the problem persists. Instead, taking time to characterize and prioritize the causes of the problem leads the team to find a permanent solution. This is where the Fishbone Diagram comes into play.

The cause-and-effect diagram is used to discover all the possible causes of a problem; those things that cause the effect; i.e., the symptoms. Further, it helps to prioritize the causes and drive a methodical identification of the true causes of a problem, thus leading to a permanent solution.

Why Cause and Effect

- You can't afford the trial and error to fix problems in your organization.
- It is a well-developed, mature process. It has been shown to be reliable and very practical.
- It taps the knowledge and experience of the entire team — a valuable source of knowledge and wisdom.
- It is user-friendly and requires no special math!

Definition

Kaoru Ishikawa is one of the pioneers of quality management processes. He was at the Kawasaki shipyards, where he became one of the founding fathers of modern management.

The Shape of Cause-and-Effect Diagrams

- A Fishbone Diagram is a methodical and graphical way to identify the root cause of a problem. It is a way to analyze symptoms, organize them and reach a logical conclusion about the cause. It is a way to plot cause and effect. The term "fishbone" is in reference to the appearance of these cause-and-effect diagrams because when completed they resemble the skeleton of fish, with the "head" being the problem statement, and the "bones" being the possible causes of the problem.
- The number of bones can vary from as few as four to as many as six, depending on the complexity of the issue or the major possible areas of the problem. Each major bone is a category of causes; for example, tools.
- Each major bone will generate sub-branches, driven by the thread of the conversation. For example, "the press does not work" may lead to causes such as "the on-off button is broken," which leads to "the contacts are burned out," which leads to "there was a power surge that burned out the contact." As you can see, there are many layers or levels depending of the depth of characterization.
- The number of layers or levels of each major bone depends on reaching the root cause. That is, when you have reached the point where you can diagnose the root cause of the problem, then you need no more layers.

The Fishbone Diagram

General Rules for Fishbone Diagrams

1. A useful Fishbone Diagram takes two to three hours to complete. Ensure that all participants have the time available. You should not rush it, as it relies on participation and brainstorming.
2. Encourage openness within your group. In some organizations, fear plays a critical role in stifling valuable input. You should reassure the participants that no one will be punished. This is called creating a safe environment. Your own behavior will have a significant impact in the degree of openness.
3. Begin by clearly stating the issue in the form of a problem. That is, you are starting with the effect and searching for the cause. For example, "The manufacturing line stopped at noon." Another example: "The phone rings too many times, and customers are upset."
4. "Why" is your best friend. Constantly asking, "Why?" will encourage the participants to better identify the causes.

The Steps

1. Always start with the effect in the form of a problem statement. For example, "There are too many rejects on station seven" or "The number of customer complaints has risen by 50 percent."
2. Encourage brainstorming to identify the cause of the problem. Reassure people that no one will be punished; it is safe to make suggestions, even if they are not 100 person certain. Encourage off-the-wall statements, and even inject humor and laughter. You must create a safe environment to enable all the information to flow freely.
3. Start by writing the input from the team members on a flip chart or marking board. Once you have enough ideas, start building the fishbone by categorizing the causes into clusters. Generally, there are four primary areas or clusters, although you may choose to have more. We discourage using more than six, as it becomes too difficult to manage. The four most common areas are equipment, people, processes and methods. For manufacturing situations, you might choose people, methods, materials and equipment. For service or business processes, you might choose equipment, policies, processes and people. Or you may replace those terms with ones that are more attuned with your culture and the nature of the problem.
4. Once you finish the categories and have exhausted all ideas, start prioritizing using your own criteria.
5. For complicated issues, you might want to select the top three or four, and have the team conduct additional research. This is particularly applicable in manufacturing or highly technical areas where validation is required.

Important

The criteria for prioritizing can be developed on the spot. For example, you might decide to use "1, 2, 3" to represent "most likely source of the problem," where 1 is the most likely. The criteria must be short and easy to understand.

Ask the team to help you prioritize using the newly developed criteria Assign responsibilities once all the prioritization is finished.

General Recommendations for Fishbone Diagrams

- Select the proper team — those people who have visibility of the problem.
- The phrasing of the problem statement is critical, as it sets the tone for the brainstorming. Carefully wordsmith the problem such that it clearly states the real issue. For example, "The press does not work" may not be sufficient. A better way might be to add more detail: "The press in station seven does not drill properly." We suggest that you work with your team to craft a statement that truly focuses on the real issue.

The Fishbone Diagram

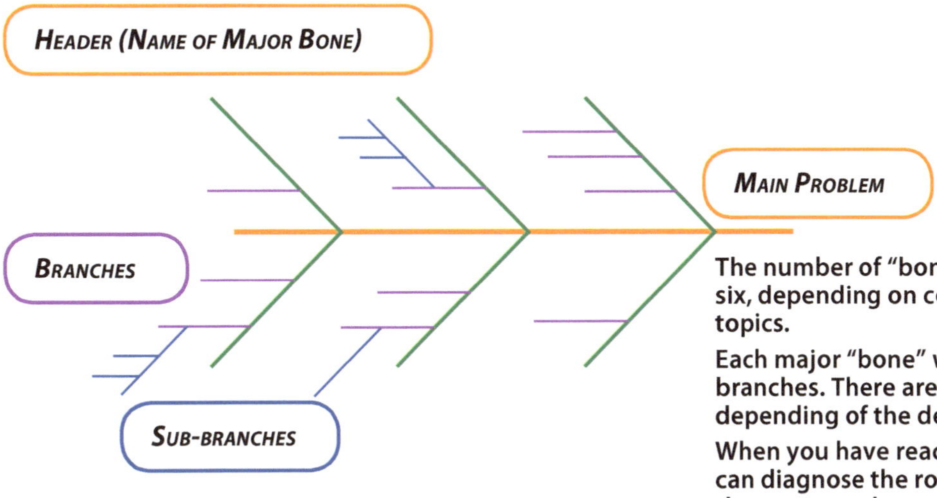

Cause-and-Effect diagrams resemble the skeleton of fish, with the "head" being the problem statement, and the "bones" being the possible causes of the problem.

The number of "bones" can vary from four to six, depending on complexity or number of topics.

Each major "bone" will generate sub-branches. There are many layers or levels depending of the depth of characterization.

When you have reached the point where you can diagnose the root cause of the problem, then you need no more layers.

Detailed Process: Fishbone Cause and Effect

1

Invite Participants

Who should attend?

- All stakeholders in the process or project. Try to keep the group to 15 participants or fewer.
- The ideal number is about 10 participants.

2

Prepare

Checklist:

- A large conference room.
- Post-It Notes® (3M brand, model #655 3x5).
- Sharpie® (Sanford brand, fine point, black markers) for all attendees.
- Flip chart paper and two easels.
- Masking tape.

3

Set up the Topic

- State the topic in the form of a problem or an issue. For example: "The technical support process does not work." Another example: "The scope of the new software application is not complete."
- Ensure that everyone understands the problem statement. You may need to rephrase it to ensure full understanding.
- Reassure everyone that there are no wrong answers.
- Reassure everyone that grammar and spelling are not important.
- Explain how to use the yellow stickies: one per subject, use the felt-tip pen, landscape orientation and NO criticism.

4

Start the Process

- Given the question or topic, ask everyone to write at least three yellow stickies outlining the reason the problem is occurring. For example, "Why did the technical support process break?" "It broke because there is no automatic answering software."
- When everyone is ready, ask each person to loudly and clearly read their yellow stickies.
- Tell everyone to listen intently as others read their yellow stickies.
- Encourage the participants to write additional yellow stickies as others present theirs.
- Collect all yellow stickies and randomly place them on the flip chart or board.
- Continue until the group runs out of ideas or repetition sets in.

Continued

The Fishbone Diagram

Cont'd

5

Start the Fishbone

- Give instructions to the group.
- Rally the group around the flip charts with the yellow stickies. The group may require encouragement to move up to the board.
- Draw the fishbone; that is, a line with four or six "ribs" emanating from it.
- Ask the participants to help you sort or cluster the yellow stickies into several topics. These topics represent each of the ribs.

6

Create the Categories (Rib Labels)

- After you have completed the sorting, label each of the ribs with a header. The header is a name that best represents why the yellow stickies in that particular rib are together.
- This is also a good time to scrub your information. For example, you can delete duplicate yellow stickies and you can rephrase or rewrite some yellow stickies to make them easier to understand.
- Important: some yellow stickies, will actually be a subset or very complementary to others. This represents a branch. So you will have the main ribs, and within each rib you may have branches. You guidance is to layer down to the root cause of the problem.

7

Prioritize

- It is best to prioritize the ribs (headers) rather than each sticky.
- Save the flip charts for archival purposes or to transcribe if it is important to do so.
- Debrief the group on the effectiveness of the tool, and ways to improve the affinity process.

See Also Developing Scales for Prioritization

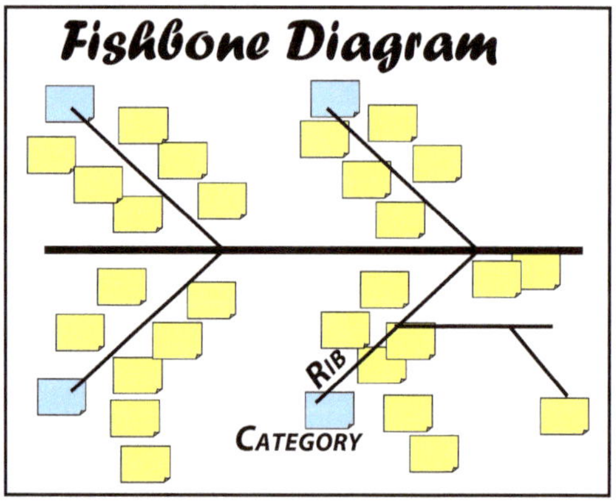

There should be from four to six ribs — don't go over six or it becomes too difficult to mange.

Example categories (rib labels):

- People
- Processes
- Methods
- Tools
- Materials
- Equipment
- Information
- Policies

8

Develop Action Plan

- Assign responsibility for additional work; for example, more research may be needed to validate your findings.
- Develop a plan to ensure that your findings translate to action that will resolve the problem.
- It may be necessary to have additional meetings to continue the process.
- You may need to form sub-teams to address specific areas.
- Above all, leave no loose ends. Ensure closure!

The Fishbone Diagram

Important The Fishbone Diagram is only a tool to diagnose the problem, specifically the root cause of a problem. It does not solve the problem. For that, you need to assign corrective action. The diagram provides the framework to methodically diagnose a problem and then diagnose the most likely cause.

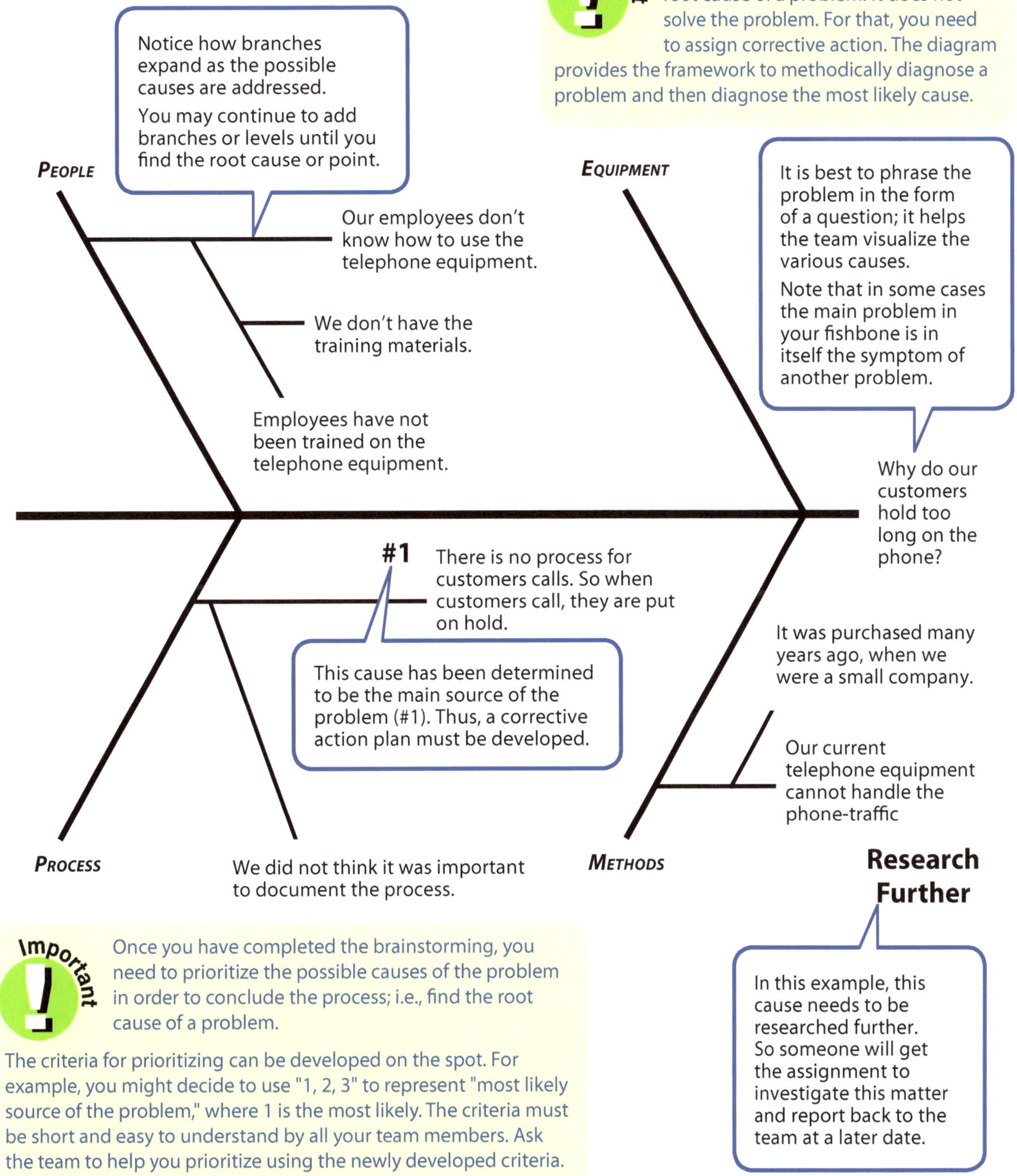

Important Once you have completed the brainstorming, you need to prioritize the possible causes of the problem in order to conclude the process; i.e., find the root cause of a problem.

The criteria for prioritizing can be developed on the spot. For example, you might decide to use "1, 2, 3" to represent "most likely source of the problem," where 1 is the most likely. The criteria must be short and easy to understand by all your team members. Ask the team to help you prioritize using the newly developed criteria. Assign responsibilities once all the prioritization is finished; for example, take all the "1's" and assign responsibilities.

The Fishbone Diagram

Developing Scales for Prioritization

Recommendations and Rules

- We highly recommend that you use simple, linear scales from 1 to 10 in every instance. This will keep the focus not on the scales, but the substance of your risk management efforts.
- Always use 1 to indicate the lowest or minimal level, while 10 will always indicate the highest or maximum level. For example, maximum damage or impact of a risk is always 10. Correspondingly, 1 will always indicate the lowest impact. It follows that the probability of a risk occurring will also be highest at a 10 and lowest at a 1. This approach will allow you to obtain information in a harmonized, or aligned, way.
- Refrain from using scales that are too complicated, such as formulas, weighted averages or algorithms. Keep in mind that this is a **QUALITATIVE** effort and adding more resolution will not add more accuracy; on the contrary, it may confuse your team.
- You and your team need to define each step in your scale — for example, 10 means Highly Likely, 8 means Likely, 6 means Moderate, 4 means Occasionally, 2 means Remote and 1 means Unlikely. You must do this in order to allow everyone in your team to have the same perception of a level. Experience will help you decide the amount of definition detail that each level requires.
- We suggest that you start by developing the description of the levels using the even numbers in your scale. This will save you time. As you become more experienced, you can complete the definitions from 1 to 10
- Be certain to document your scales and the definition of the levels. This will allow for repeatability, reference and continuous improvement.

Example: Simple Scales and Definition Levels

A hypothetical scale for probability might describe the levels as shown in the figure below:

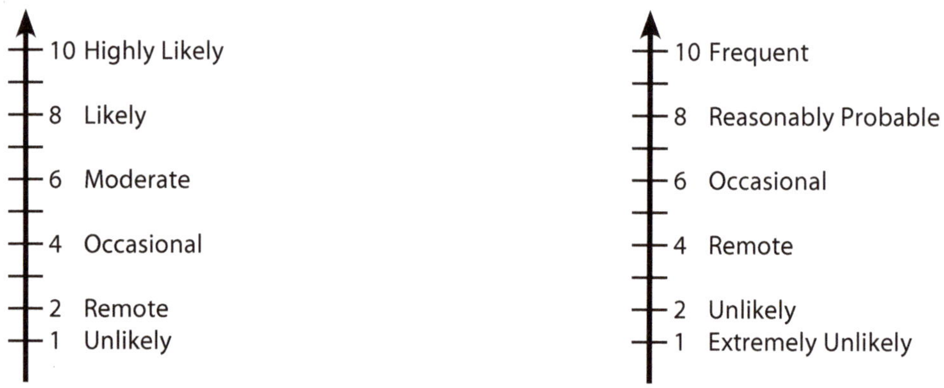

Here are two very simple definitions of a scale for the probability a risk might occur. The meaning of each level (step) of the scales would have been determined by the consensus of the team. You and your team could use either scale as a starting point.

Over time, you can add additional definitions and descriptions among the different levels. Your team will become familiar with the meaning of each number, and the uncertainty between levels will be reduced.

Example: Sophisticated Scale and Definition Levels

A hypothetical scale for impact or damage might describe the levels as shown in the table below: Over time, you can develop scales that are much more meaningful to your situation.

Level	Rank	Severity of Failure as Seen by the Customer
1	None	No noticeable defects.
2	Very minor	Abnormal noises, noticeable only by the most discerning operators.
3	Minor	Abnormal noises, noticeable by 50% of the customers. Fully operational, but causes stress on the operator.
4	Very low	Abnormal noises, noticeable by the customer. Fully operational, but causes stress on the operator.
5	Low	Operable but difficult to operate, all functions available.
6	Moderate	Operable but difficult to operate, some functions not available.
7	High	Operable, but at reduced level of performance.
8	Very high	Device inoperable.
9	Hazardous with warning	Unsafe to operate.
10	Hazardous without warning	Operator injury very likely.

Here is an example of a more sophisticated definition of a scale, one that focuses on the levels of impact for each risk. The meaning of each level (step) of the scales would have been determined by the consensus of the team.

Process Mapping

NEW TOPIC

Steps to Success

A process is a **DOCUMENTED**, sequential list of **VALUE-ADDED** steps, the aggregate of which is used to deliver a product to a **CUSTOMER**.

- **DOCUMENTED:** A process must be documented; otherwise, it fails the most critical test. Furthermore, documenting the process has several advantages. It forces the dialog to be collaborative. It allows more than one person to study and use the process. Also, it's the only way to initiate formal improvements.
- **VALUE-ADDED:** Every step in the process must add value, as defined by the customer, or it is waste that must be removed.
- **CUSTOMER:** Every process must have a customer. For example, a customer of Marketing is Engineering. Thus, Marketing is the Process Owner, responsible for creating the product requirements, and Engineering is the Process Customer. Engineering has a set of needs, and it is the responsibility of Marketing to identify and satisfy these needs.

Create a Process Map

Creating of a process is not difficult. You can start by reviewing the many process maps shown in this guidebook. Collaborate with your peers, and define a process using the Yellow Sticky Protocol.

1. Call a meeting with people closely involved in the process you have chosen to document. For the process of producing a new brochure; for example, you would likely invite marketing staff, your advertising manager (marcom) and your advertising agency if you have one. These meetings generally last from two to three hours, sometimes longer.
2. Start the meeting by defining the process you wish to document — in this case, the creation of a new brochure. Also define the start and end of the process. Your team will help you.
3. Define the customer. In our example, the process customer happens to be your company's external customers. Be careful to differentiate between customer and stakeholder. In our example, the marketing manager is not a customer, but a stakeholder with a vested interest in making this process work. Furthermore, your general manager is also a stakeholder.
4. Define the output (result or product) of using the process. In our example, it may be "a printed brochure, ready for delivery to our customers." This critical step clearly defines the endpoint.
5. Brainstorm with your team to identify all the steps. Using yellow stickies, one for every step. At this point, you don't need to be concerned with value, sequence, duration, etc. of the steps.
6. Organize the steps sequentially, from process start to end. Add and refine steps as needed. Eliminate redundant or trivial steps. Take your time; this adds much clarity to your process map.
7. Study your map to ensure accuracy and clarity. Is the sequence of steps the way things really happen? Are there any steps that don't make sense? Are the steps described in a way that's easy to understand?
8. Test every step against the definition of the process. Pay particular attention to the value-added criteria. Ask yourself, is this step necessary? Does it really add value? This is your first opportunity to improve efficiency.
9. Finally, review your process map one last time, document it for permanent use and start using it.

Continuous Improvement

Once you begin using your documented process, the payoff begins. But the payoff continues as you make continues improvements to your process map. As you use it you will discover obvious ways to improve it — for example, redefine, reorder or eliminate steps. The secret is to constantly review your process map to identify opportunities for improvement.

Appendixes

PRODUCT DEVELOPMENT

Recommended Books

APPENDIX A

Introduction

These lists reference books that the senior consultants at Rapidinnovation, LLC have read and found applicable to today's business challenges. Rapidinnovation, LLC is a consulting firm whose primary focus is serving product development organizations around the world, which we've been doing since 1997, and with many satisfied clients.

We welcome your recommendations and reviews of these and other books and resources. Please email the authors with your comments and suggestions.

Teams and Teamwork for Product Development

Book Title & Author	Subject	Comments
The Wisdom of Teams By Jon R. Katzenback Published by Harvard Business School Press	• High-performance teams.	• Excellent book for creating development teams. • Clear, no-nonsense approach to cross-functional teams such as concurrent engineering, product development, etc.
Peopleware, Productive Projects and Teams, Second Ed. By Tom DeMarco & Timothy Lister Published by Dorset House Publishing	• Software development. • People management.	• Good book to help understand the human side of software development.
Principle-Centered Leadership By Stephen R. Covey Published by Fireside	• Timeless principles of leadership. • Clear, no-nonsense understanding of leadership in a business environment.	• The difference between a manager and a leader, and its implications in an organization.
Punished by Rewards By Alfie Kohn Published by Houghton Mifflin Company	• Reward, recognition, celebrations and motivation in organizations.	• The truth about failed recognition programs and the unintended results of poor motivation and recognition programs.
Mastering Virtual Teams, Third Ed. By Debora L. Duarte and Nancy Tennant Published by Jossey-Bass	• The ways and means for enabling virtual teams to succeed.	• Duarte does a good job of presenting the challenges and solutions for distributed teams. A level-headed presentation of the issue.

V-BOOKS/011912/3.0.1

Recommended Books

General Management and Strategy

Book Title & Author	Subject	Comments
The Innovator's Dilemma By Clayton M. Christensen Published by HarperBusiness	• Why companies fail to stay at the top of their industries.	• A book that confronts a reality faced by most successful technology companies: the need to change, and change fast.
The Discipline of Market Leaders By Michael Treacy Published by Perseus Books	• Choose your customers, narrow your focus, dominate your market.	• Deliberately choosing the customers you wish to serve allows you to develop the competencies needed to succeed.
Built to Last By James C. Collins Published by HarperBusiness	• Visionary companies. • Values, culture and successful businesses.	• Focusses on the human side of corporate culture for the promotion of long-term success.
The Balanced Score Card By Robert S. Kaplan and David P. Norton Published by Harvard Business Press	• What to measure and why. • Financial measures alone are not enough to ensure sustainable customer satisfaction and profits.	• Rational approach to measures. • Systemic look at the enterprise. • Focus on the human side of measures.
Companies with a Conscience, Third Ed. By Mary Scott and Howard Rothman Published by Citadel Press	• Case studies of 12 companies that are profitable and at the same time people-, environment- and community-friendly	• Good examples of "having your cake and it eating it too" — the ability to make a profit and have a clear corporate conscience.

Product Development

Book Title & Author	Subject	Comments
Rapid Development: Taming Wild Software Schedules By Steve McConnell Published by Microsoft Press	• Perspectives and methods for rapid software development. • Focus on the process and framework for development.	• The methodical approach to rapid software development. • Good food for thought.
Revolutionizing Product Development By Steven C. Wheelwright Published by The Free Press	• One of the classics on product development.	• A must-have for those who are serious about product development.
The Invisible Computer By Donald A. Norman Published by The MIT Press	• Why good products can fail.	• Insightful as to why technology companies behave like technology companies.
Dynamics of Software Development By Jim McCarthy Published by Microsoft Press	• Rapid software development is not the result of countless hours of coding.	• A book more about teams and teamwork than software development.

Recommended Books

Book Title & Author	Subject	Comments
Portfolio Management for New Products, Second Ed. By Robert G. Cooper, Scott J. Edgett and Elko J. Kleinschmidt Published by Perseus Books	• Managing of technology for profits. • Disciplined product development. • Separate the losers from the winners.	• A solid, direct book on managing a portfolio of products while optimizing resources. • Good methods for balancing your product offering.

Productivity Improvement

Book Title & Author	Subject	Comments
The Fifth Discipline By Peter M. Senge Published by Doubleday	• System thinking. • The quintessential book of systemic approaches to solve business problems.	• A must-read to understand the total solution to customers' problems • Also very important to re-engineering
Improving Performance: How to Manage the White Space on the Organization Chart By Geary A. Rummler Published by Jossey-Bass	• Business processes. • Teamwork in organizations. • Process mapping for non-manufacturing organizations.	• Clear understanding of processes in a business environment. • Good how-to information and a level-headed approach to organizational development.
Kaizen: The Key to Japan's Competitive Success By Masaaki Imai Published by McGraw Hill	• Continuous improvement.	• The truth about continuous improvement as a culture and not a fad.

Recommended Books

Project and Program Management

Book Title & Author	Subject	Comments
Flexible Project Management for Product Development By Jose Campos Published by Rapidinnovation, LLC	• Clear understanding of the conventional project management rules and tools do not necessarily work in a product development environment. But here, the author has experience in both project management and product development.	• Easy to understand and very detailed. • Plenty of examples, templates and forms ready to use. • Easy to share with other members of the development team. • Clearly written with product development in mind, very practical.
Project Management Toolbox By Dragan Milosevic Published by John Wiley and Sons	• A comprehensive set of project management tools full explanations and plenty of examples	• A must-have reference
Program Management for Improved Business Results By Dragan Milosevic, Russ Martinelli and James Waddell Published by John Wiley and Sons	• A business approach to the role of program managers. Authored by Dragan Milosevic, acclaimed authority in project and program management.	• Probably the best book on program management.
Reinventing Project Management By: Aaron J. Shenhar Published by Harvard Business Press	• Proof positive that every project is different and therefore, you cannot manage them exactly the same.	
The Fast Forward MBA in Project Management, 2008 Ed. By Eric Verzuh Published by John Wiley and Sons	• Looks at the challenges of today's project environment and examines the skills a project leader needs to make a project successful. • It includes organizational changes, tools and techniques, control methods and real-life applications.	• Good medium-depth overview of all aspects of project management. • Easy to read, with examples and case studies, but meaty enough that both junior and experienced project leaders will find nuggets. • Down-to-earth
The New Project Management, Second Ed. By J. Davidson Frame Published by Jossey-Bass	• Frame makes a case that conventional wisdom is only marginally relevant to project leaders working in today's turbulent project environments. • He offers strategies and techniques for changing from implementor to initiator, and for dealing with the soft and complex issues that affect most technology projects.	• Good ideas, especially on soft skills and orientation toward the customer. • Frame sometimes undervalues the traditional techniques and gets too enthusiastic about how empowered a project leader can be. Pair this book with a solid grounding in the hard fundamentals of project management, such as from Verzuh's book.

Recommended Books

Book Title & Author	Subject	Comments
Visualizing Project Management, Third Ed. By Forssberg, Mooz and Cotterman Published by John Wiley and Sons	• A recently updated project management textbook with lots of examples. It strives to provide an integrated model of project management, and focuses on vocabulary, teamwork, the project cycle and management techniques.	• Covers a wide variety of hard and soft project management topics, with short case studies of how they have been used in real situations. • Focuses mostly on applying them within a sequential framework, so don't look for lots of insight on iterative approaches here. The authors' experience in aerospace and government projects shows through in the vocabulary and examples.
The Wiley Guide to Managing Projects By Peter W. G. Morris and Jeffrey K. Pinto Published by John Wiley and Sons	• Published in 2005, this book is up-dated every 10 years. This is a comprehensive book on project management. Looks well beyond the basics on project management (cost, schedule and performance).	• A must-have reference for those who are serious about project management. • Critically important for those involved in the strategies of project management.

The Voice of the Customer (VOC)

Book Title & Author	Subject	Comments
The Voice of the Customer for Product Development By Jose Campos, Jean-Claude Balland Ph.D. Published by Rapidinnovation, LLC	• A detailed guidebook on how to capture and use the voice of the customers (VOC) for product development. The author describes every step and detail in an easy to understand fashion.	• Easy to understand and very detailed. • Plenty of examples, templates and forms ready to use. • Easy to share with other members of the development team.
How Customers Think By Gerald Zaltman Published by Harvard Business School	• Essential insight into the mind of the customers and the market.	• Customer needs are much more subjective than we thought. • An eye opener concerning value.
Advanced QFD By Larry Shillito Published by Wiley & Sons, Inc.	• The use of QFD (Quality Function Deployment) in a real-world environment, linking technology to the market.	• A sound, levelheaded book on QFD, A must-have if you are serious about QFD.
Software Requirements, Second Ed. and More About Software Requirements By Karl Wiegers, Published by Microsoft Press	• A practical guide for gathering and managing requirements throughout the product development cycle.	• As relevant to product requirements as it is to software requirements, this practical guide will go a long way towards making your R&D and Marketing people see the world through the same lens.

Recommended Books

Book Title & Author	Subject	Comments
Customer-centric Product Definition By Sheila Mello Published by AMACOM	• Good coverage of the Voice of the Customer (VOC) and its use in defining new products.	• Very good book; the author clearly portrays the process of getting the customers at the center of new product development.
Writing Effective Use Cases By Alistair Cockburn Published by Addison Wesley	• A practical guide to the powerful approach of expressing the behavioral requirements of complex systems.	• Practical, with numerous examples of good and bad use cases.
Customer Visits, Second Ed. By Edward F. McQuarrie Published by Sage Publications	• A clear description of what it takes to perform a good customer visit.	• The author is well-recognized as an authority in this field. Levelheaded approach.
Acquiring, Processing, and Deploying the Voice of the Customer By M. Larry Shillito Published by St. Lucie Press	• Description of ways to obtain, process and use the voice of the customer.	• Plain English, down-to-earth, but clearly from someone who knows.
Voices Into Choices By Gary Burchill Published by Joiner	• A handbook with step-by-step descriptions to obtain and use the voice of the customer	• Comprehensive. • Truly a fine work — a must-have handbook.
Priceless: Turning Ordinary Products Into Extraordinary Experience By D. LaSalle & T. A. Britton Published by Harvard Business Press, 2003	• The title says it all. People don't buy products, they buy value and experience. This book will help you see the world through the eyes and heart of customers.	• Very valuable in bringing the new concept of experience to a tactical level.
Quality Function Deployment: How to Make QFD Work for You By Lou Cohen Published by Addison-Wesley, 1995	• A tour of QFD fundamentals, as well as advanced concepts	• Clear exposé to a sophisticated subject. • Most useful to engineers than marketers.
The Observational Research Handbook: Understanding How Consumers Live with Your Product By Bill Abrams Published by McGraw-Hill Professional, 2000	• Practical coverage of the art of observing customers in context to understanding latent needs.	• Useful advice on how and when to observe customers to get clues to their needs.
Learning from Strangers: The Art and Method of Qualitative Interview Studies By Robert S. Weiss Published by Simon & Schuster, 1995	• The subtitle of the book says it all: "The art and method of qualitative interview studies." These are important skills to master.	• A how-to book on interviewing customers by an expert on the topic.

Recommended Books

Risk Management and FMEA (FMCA)

Book Title & Author	Subject	Comments
Risk Management and FMEA for Product Development By: Jose Campos Published by Rapidinnovation. LLC	• A detailed guidebook on risk management for product development. The author describes every step and detail in an easy to understand fashion.	• Easy to understand and very detailed. • Plenty of examples, templates and forms ready to use.
Waltzing with Bears: Managing Risk on Software Projects By Tom DeMarco and Timothy Lister Published by Dorset House Publishing		
Proactive Risk Management: Controlling Uncertainty in Product Development By Preston G. Smith and Guy M. Merritt Published by Productivity Press	• A "plain English" book on risk management.	• A good reference for anyone responsible for managing risk.
Managing the Unknown By Christoph H. Lock, Arnound DeMeyer and Michael T. Pich Published by John Wiley & Son	• A book about managing the risks you don't know: the ways and means to deal with total uncertainty. • Intriguing new concepts about risk, particularly in areas with high uncertainty such as new technology.	• A good book to have if you are in product development or technology development. • Challenges conventional views of risk management.
Project and Program Risk Management: A Guide to Managing Project Risks and Opportunities By R. Max Widerman Published by Project Management Institute, 1992	• A risk management book for beginners and intermediate managers of risk.	
Failure Mode Effect Analysis: FMEA from Theory to Execution, Second Ed. By D. H. Stamatis Published by American Society for Quality, 2003		• A comprehensive book on FMEA, strong in context and advanced methods. • A good overall reference book for those serious about FMEA.
Project Management Tool Box By Dragan Milosevic Published by John Wiley and Sons, 2003	• A comprehensive set of project management tools, full explanations and plenty of examples.	• A must-have reference.

Glossary

Definitions for Product Development

Practical Definitions in Plain English to Enable Effective Action

The intent of this glossary is to explain the terms you will encounter in the real world of product development. Furthermore, we explain these terms in a practical, common-sense fashion. When multiple relevant definitions exist, we will provide them. With this resource, you will be able to better connect the concepts and tools of product development to your projects.

Term	Definition & Notes
Agile	• A software development model or framework based on frequent iterations with the customer and development of the software in small, incremental amounts, collaboration and flexibility.
Alpha Unit	• A version of a new product that is ready for internal testing. The product meets criteria that enable testing that will yield valuable data and input to proceed with the final version. • After the Alpha Unit comes the Beta Unit, a more mature version of the product. • In hardware an alpha unit may be used to test the manufacturing process, functionality or testing strategy. • In software it can be used to test human interface (GUI), usability or functionality. • Alpha units are not saleable and are intended for internal use only.
Assumption(s)	• A short set of relevant factors that can have a positive or negative impact on your project or program and define the playing field. – The playing field refers to the arena or area of focus for your endeavor. – A short set means three to seven relevant factors. – Relevant means of importance to you and your team — that which must be taken into consideration when making decisions.
Attribute	• Used to define a set of general characteristics of a platform, product line or product, an aggregate of several features. Examples include "ease of use," "ruggedness" and "re-configurable". • Related to, but not the same as "Product feature." See also "Feature."
Awareness	• A measure of the percent of target customers who are aware of the existence of something. It could be a product, a brand or even a concept. • Awareness does not imply preference. • Awareness can be measured for brand, product or any other attribute you select. • Awareness should be measured before and after an event; for example, an advertising campaign. The difference provides a measure of effectiveness. • There are two kinds of awareness: Unaided awareness is when the customer volunteers the information (manifests the awareness) without any help. Aided awareness is when a hint or suggestion is provided, such as a photograph, and then the customer expresses awareness.

Glossary

Term	Definition & Notes
Benchmarking	• The process of studying how successful companies, industries and even competitors perform specific functions and then selecting the best as standards of performance for your own improvements. • You can benchmark with your competitors or across industries, in your own country or across the world. • Benchmarking is a good way to develop your level of excellence, and to learn "good practices" for your continuous improvement.
Best Practices	• A method, technique, tool or other practice that has shown to be the best within a certain category. Best practices must be researched, defensible and published by an authorized organization. • Methods, approaches or processes deemed to be "best in class"and, consequently, worthy of imitation. • See also "Good Practices."
Beta Test (a.k.a. Beta Unit or Beta Version)	• An external test of pre-production products. A way to test new products in actual application to discover bugs and last minute fixes before full introduction. • Some companies (software in particular) use Beta Tests as part of their introduction strategy, to generate awareness and word of mouth.
Bill of Material (BOM)	• A hierarchical list of systems, sub systems, sub assemblies and components that make up a product or a higher level sub-system. • It can be a set of parts list, drawings, and schematics. In manufacturing it can also mean the way a new product is assembled including all parts, systems and sub systems.
Brainstorming	• The process of tapping into a group's creativity to generate ideas, solutions or innovation. • Generally, it takes the form of a meeting or gathering where the team or group is encouraged to come up with as many creative ideas as possible. • There are many methods to facilitate or enable brainstorming. The methods are intended to enable the ideation, processing of the information, and finally prioritization of the ideas. Many brainstorming sessions fail due to a lack of a framework or process.
Brand	• The combination of all tangibles, intangibles and actions of a company that create a particular promise of value in the customers' mind. • A brand is a holistic concept that evokes the mental images and expectations the customers have about a particular company.
Breakeven	• A financial metric that identifies the point where the revenue (sales) of a new product match the investment made in developing it. • At this point, the new product has reached enough sales to start generating "positive cash flow."

Glossary

Term	Definition & Notes
Business Case	• The results of the business analysis to demonstrate the viability of a new product or service prior to approval for development. • Viability includes profitability, market share or other critical and strategic objectives of your company. • "Making the business case" in product development implies your ability to convince the company executives that investing in your new product will result in a measurable return. Philosophically, you are going to the company executives to ask for an investment, which will results is a quantifiable return
Business Model	• The ways and means selected by the company to generate revenue and make a profit. • The Business Model is said to be working when the company has deployed and executed the strategies to generate revenue, profits and serve its customers.
Business to Business (B-to-B)	• Sales to other businesses and not to individual consumers. For example, an OEM sells components to a manufacturing company. • Sales to distributors, representatives or wholesalers are also considered B-to-B.
Champion	• A person who takes strong personal interest in seeing that a particular process, product or other is carried to fruition. Champions are always ready to deal with adversity and committed to success. • Champions are said to be "maniacal" in their commitment, willing to put exceptional effort and energy into their cause. • No innovation can be brought to fruition without champions.
Change Management	• An organizational process aimed at helping employees embrace changes in their current business environment and then change their behavior in support of the new state. • A structured approach to shifting or transitioning individuals, teams and organizations from a current state of behavior to a desired future state.
Charter, Team Charter	• The document that officially starts an initiative. This document outlines the team members, the common purpose that brings them together and the performance goals to ensure progress and outcomes. It also defines the roles, responsibilities and authority. It defines the desired approach to ensure completion. • The team members should be strongly involved in — even responsible for — the development of the charter. • The Charter must be fully supported by Management. • A charter is the unifying principle of the team. Without it, each team member has a different idea of what needs to get done and how.
Co-location	• Product development where the entire development team is in one location; for example, the same building. • The processes and methods for co-location development are different from that used for multi-site or distributed product development.
Commercialization	• The process of taking a new product to market. The transformation of an idea into a product that customers are willing to purchase. • Commercialization can be used for a new product, a new technology or anything that a particular set of customers value and are willing to pay to obtain it.

Glossary

Term	Definition & Notes
Compelling Reason To Act (CRTA)	• The needs the customer must address now so that a budget and the will to act can be generated. • CRTA is proposed to designate the highest priority needs — the ones that can be funded and will be acted upon by your customer. • A customer must have a CRTA in order to purchase your product. It is your responsibility to identify CRTAs so you can develop products that address customer requirements.
Competency	• The aggregate set of skills, knowledge and abilities that an individual needs to fulfill a specific set of measurable responsibilities. • Competency is all about one individual — not a team or an organization. The equivalent statement at the organizational or team level is known as "maturity" or "organizational maturity"
Complete Product	• The aggregate of the tangible, intangible and emotional components that constitute a product in the eyes of the customer. • See also "Offering" and "Whole Product."
Completion	• The verifiable fulfillment of all objectives. • The conclusion of all efforts; for example, the completion of a new product development is the introduction or rollout to the market place.
Compound Annual Growth Rate (CARG)	• The calculation of rate at which an investment would grow if it grew at a steady rate. It's a way to smooth the return on an investment for the purposes of making informed decisions. • In product development, CARG is generally used to describe the growth rate of a market or the growth rate of new products.
Computer Aided Design (CAD)	• The use of computers and software to design a new product or parts of a new product. • CAD is a very advanced technology enabling simulation, testing, and all aspects of design.
Computer Aided Manufacturing (CAM)	• The use of computers and software to design and run a manufacturing process in an integrated fashion; that is, as a system. • In some areas CAM is also used to describe any aspect of the manufacturing process that is run with the aid of computers or fully automatic, not necessarily the entire system of manufacturing. • Widely used in industry, CAM has become essential to running manufacturing operations around the world. • CIM (Computer Integrated Manufacturing) may be used in conjunction with CAM.
Computer Integrated Manufacturing (CIM)	• See "Computer Aided Manufacturing (CAM)"
Concurrency	• The ability to develop several products at the same time. • The goal of a product company is to achieve speed, concurrency and repeatability. • It can also be applied to the development of a single product where parts, assemblies and subassemblies are developed at the same time and not sequentially.

Glossary

Term	Definition & Notes
Concurrent Engineering	• A model of product development where all functions such as engineering, marketing, manufacturing, etc., work at the same time to design and take the product to market. • A set of methodologies that enable rapid product development by allowing as many activities as possible to take place simultaneously. • Contrasted to the sequential model, where one activity had to wait for another before it could be started.
Concurrent Engineering	• An emerging engineering methodology based on the premises: – All aspects of the life-cycle of a product should be considered as part of the initial design process; that is, from basic functionality until end of life. – All activities that can happen at the same time must be undertaken. • The methodology, if properly applied, yields better products in a shorter amount of time.
Conjoint Analysis	• A quantitative market research technique used to characterize how people value the features (or benefits) of a product. • One of the few methods available to determine the possible price of a new product with objectivity. • Conjoint analysis is powerful because it puts the customer in a real life situation of having to make trade-offs between benefits (what he wants to get) and costs (what he has to give to get it).
Constraint	• Restriction or boundary imposed on a project or design. • Constraints may help in setting priorities, avoiding risks or the physical limitation of the technology.
Contextual Inquiry	• A way to gain knowledge about a customer by observing people in their environment doing specific activities. • The use of anthropology, social science and other techniques to study customers performing specific tasks for the purpose of understanding their needs. • The observation must be qualitative.
Core Competency	• The few things that a company does that: – Its customers find exceptionally valuable. – It does better than its competitors. – Its competitors find difficult to imitate. – It would never outsource to anyone. – It can transfer to other industries as is. – Is part of its growth plan. • The definition of Core Competency varies. It's driven by the specific situation of a company at the time it defines its Core Competencies. Nonetheless, it is critical for a company to understand and develop its core competencies. • Core competencies are always a web of things that a company does particularly well without sometimes even be conscious of it.

Glossary

Term	Definition & Notes
Core Team: Concurrent Engineering	• A small group of representatives from the critical functions (e.g., Engineering, Marketing, Program Management, Finance, etc.) responsible for the development of a specific product and accountable to deliver measurable results. • The Core Team is responsible and accountable for progress and eventual delivery of the new product to the market. It acts as the Program Manger for the development.
Core Team: Generic	• A small group of three to five individuals that is chartered and empowered to guide the implementation and completion of an initiative. • Very useful in fast-changing environments where distractions abound because a small group can focus attention and energy to ensure clean execution.
Corrective Action	• The specific steps to eliminate the cause of a detected nonconformity or other undesirable situation. That it, the steps that are taken to permanently remove the causes of an existing nonconformity through root cause analysis and deploying permanent corrections. • In contrast, "Remedial Action"refers to the temporary steps taken to eliminate one specific non-conformance without regards to a permanent solution.
Criteria	• A standard used to evaluate alternatives in a decision. Criteria are always finite and easy to understand by all. • In the product development process, the criteria are generally defined by the Gatekeeper at major gates.
Critical Path (CP)	• The path with the longest duration • CP comes from the analysis of a network diagram or a PERT (Project Evaluation and Review Technique) diagram in project management.
Critical Success Factor (CSF)	• Something that is critical to your program and must happen. • The few key areas where things must go right for the business to flourish. If results in these areas are not adequate, the organization's efforts for the period will be less than desired. • Those few things that must happen in order to achieve the desired results.
Cross Section (Market)	• A typical selection or sample. For example, a cross section of a particular market segment. • This term is a metaphor from engineering, used to reflect the framework of a particular structure such as a market segment.
Cross-Functional Team (a.k.a Multi-Functional Team)	• A team made up of representatives from various functions. For example, marketing, engineering and manufacturing. • Most product development is achieved though cross-functional teams.

Glossary

Term	Definition & Notes
Culture	• A mosaic of interrelated elements and their collective day-to-day interaction that shape the behaviors of people in the group and what's right and wrong. • For example: – Management behavior. – Customs, myths and norms. – Ceremonies, rituals and events. – Rewards and consequences. – Physical environment. – Rules and policies. • A company culture develops when corporate values are consistently followed for a number of years. • Said another way: "Culture Consists Of A Shared, Commonly Held Body Of General Beliefs And Values That Define The 'Shoulds' And The 'Oughts' Of Life." — Kluckhohn
Customer	• The person or entity that pays for the value of a product or service. • There is tendency in some organizations to refer to teams and functional groups as "internal customers." This tends to diffuse the importance of the one and only true customer, the one that pays. • Customers can be "users" or not of your product. Make the distinction between the person who pays and the person who uses.
Customer Experience	• A succession of events in the life of the customer dealing with an offering. This interaction will generate good or bad feelings. • Designing experiences is the new frontier in user-centered design because creating pleasant feelings is the best way to generate satisfaction, loyalty and word of mouth. • Thinking in terms of customer experience forces the marketer to consider the entire sequence of events that the target customer undergoes when dealing with a company and its offerings.
Customer Intimacy	• The business definition as introduced by Michael Treacy in Discipline of Market Leaders is (paraphrase): The discipline of delivering what each specific customer wants or needs rather than what the market as a whole wants or needs. • Customer-intimate companies develop deep relationships with their customers so as to anticipate their needs and solve their problems better and faster.
Customer Requirements	• The crystal-clear articulation of the desired outcomes a customer is striving to achieve.
Customer Satisfaction	• The positive or negative evaluation or feeling of customers that results from comparing the actual experience with a product or service with the expectations it had prior to purchase and use. • Customer satisfaction is a different concept than "customer value." A customer may be satisfied or not with the value it gets.
Customer Value	• The perception of what customers will get (benefits) for what they have to give (costs). • Alternatively, it is the customers' perception of the net benefits they will derive from your offering at a particular price, situation and use.

Glossary

Term	Definition & Notes
Customer Value Drivers (CVDs)	• The criteria that are perceived by the customer to be the most important to the buying process. The yard stick that the customer uses to make the buy decision. • Customer value drivers can be tangible (objective) or intangible (subjective).
Customer-Driven	• The urge to grow a business culture committed to the relentless creation of superior value for the customer. • It is a set of values that put the customers' interest first in order to develop a long-term profitable enterprise — while not excluding those of other stakeholders (paraphrased from: Narver and Slater). • In a for-profit business it is part of the overall strategy to generate revenue and profits. A company can make a profit and not be customer-driven.
Defect	• Lack of conformance to a standard. • Since the introduction of Quality methods, a defect is anything that your customer says it is. This is an effort to focus the organization to evaluate products and services as the customer would.
Defects: Engineering	• Engineering "defects" include design errors, software bugs, incorrect documentation and any substandard design or development steps that would not occur if the process were under control.
Deliverables	• The visible or "touchable" results from a task or a project. The outcomes from a task or a project. • Deliverable has come to mean more than "able to be delivered", now more at "the results of an endeavor or a project." • For example, the deliverable from the marketing organizations is a Marketing Requirements Document (MRD).
Delphi Process	• A technique that uses iterative rounds of consensus development across a group of experts to arrive at the most probable outcome for some future state. • "Experts" are people who have knowledge in a particular area; for example, knowledge about a specific market segment.
Derivative Product	• A new product based on changes to an existing product without affecting the basic product architecture or platform. • A derivative product can be an improvement to an existing product or a different version to satisfy a specific set of customers. • In software development, a rev includes new value to the customer. A rev to fix bugs is not a derivative product.
Design Change Order (DCO)	• A process to ensure the smooth and controlled change of a design and minimizing disruption. • Documented changes to the design of a new process, which is part of a formal process for changes. • The purpose of the DCOs is to minimize disruptions to the design process and ensure clarity of communication. • Also used in this context: – ECN: Engineering Change Notice – ECC: Engineering Change Control – ECO: Engineering Change Order

Glossary

Term	Definition & Notes
Design Engineer	• A person with proven domain expertise and experience who is responsible for creating innovative solutions that customers value. • The output of a design engineer is a package of technology in the form of documentation (drawings) that is the tangible manifestation of an innovative solution that customers value.
Differentiated Product	• Characterizes a product that provides more advantages than its competitors for the most important Customer Value Drivers of a given segment. • A well-differentiated product is one of the most important success factor for a new product.
DOA (Dead on Arrival)	• Products that arrive at a customer site unusable. • A product that does not work when the customer unpacks it.
Early Adopters	• Customers who buy into a new product or technology very early in its life cycle. • Early Adopters are willing to take the risk of an unproven new product or technology to gain a business or personal advantage.
Economic Value Analyses (EVA)	• A methodology that enables you to understand the real progress of a development process by measuring actual progress towards completion. A way to clearly understand progress towards the objectives of a new product. • EVA is found in project management texts — very popular in the construction industry — and now being adopted in product development due to its objectivity. • While budge and schedule are indicators of "movement", they do not necessarily show progress towards the achievement of the objectives of the new product.
Elevator Pitch (a.k.a. 30-Sec. Commercial)	• A short statement summarizing in a compelling way a value proposition. • If a pitch can't interest a prospect in the time it takes to move a few floors in an elevator, you don't have a strong, well-differentiated offering. Rework the pitch and try again.
Engineering Change Control (ECC), **Engineering Change Notice (ECN),** **Engineering Change Order (ECO)**	• A formal process that documents a needed design change regardless of the root cause of such change. The process must follow sound engineering principles for control, tracking and disposition. • See "Design Change Order (DOC)." • EISs may include modeling, simulation, analysis and other computational requirements.
Engineering Information System (EIS)	• Applications and solutions for information-intensive engineering and scientific endeavors. The ability to model, process and manage very large amounts of information for the purpose of enabling engineers and scientists to accomplish their goals.
Experience	• A succession of events in the life of the customer dealing with an offering. This interaction will generate good or bad feelings. • Designing experiences is the new frontier in user-centered design because creating pleasant feelings is the best way to generate satisfaction, loyalty and word of mouth. • Thinking in term of customer experience forces the marketer to consider the entire sequence of events that the target customer undergoes when dealing with a company and its offerings.

Glossary

Term	Definition & Notes
Failure	• A defect, nonperformance or lack of adherence to a specific requirement. • Something that causes a product to no longer fulfill its intended function.
Failure Analysis	• A process to methodically characterize and determine the root cause of a failure. The reasoning from the symptoms of a failure to the identification of the root cause of the failure. • Used in reliability engineering, as well as FMEA (Failure Mode and Effects Analysis).
Feature Creep	• The tendency to add new features and attributes to a new product during the design process. • Feature Creep is not a desirable activity as it delays the product, adds to its complexity and may even interfere with the business objectives of the new product. It also increases the stress and aggravation of the entire development team.
Feature (a.k.a Attribute)	• A quality or capability of an offering. • When they are familiar with the product category, customers often express their needs in term of features or attributes (e.g., "I need a 4-door sedan that gets more than 25 mpg") making it more difficult to understand their primary needs.
Firefighting	• The act of addressing too many interruptions, trivial issues and other distraction at the expense of the schedule. • Taking care of surprises and other unexpected distractions. • Firefighting is generally caused by a severe lack of planning and the lack of a risk management plan.
FMEA (Failure Mode and Effects Analysis)	• A structured process to identify potential failures of new products, components or subsystems. A form of risk management. • Can also be used to analyze and determine the root cause of actual failures.
Focus Groups	• A market research tool where a qualified facilitator (moderator) leads a discussion with a select group of target customers for the purpose of understanding their needs and preferences. • Focus groups are generally eight to twelve customers in a single room; generally the room is designed for focus groups. The facilitator (moderator) leads the conversation, and the response is observed and also video taped for later analysis. • Focus groups are a qualitative form of market research.
Functional Test	• Testing a circuit board to identify functional level faults or defects.
Funnel	• See "Roadmap."

For comments, questions or additional copies please e-mail the author: flexpm@rapidinnovation.com.

Glossary

Term	Definition & Notes
Fuzzy Front End	• The beginning of a program or project when much is not known and a high level of confusion exists. It's also the best opportunity to influence the long term success of the program or project. All programs and projects have this phase. • Initially, this term was defined as the period of time when the development of a new technology was not clear (fuzzy). The meaning has been extended to include the start of a product development project when much confusion exists — more questions than answers. • We suggest that every development project has a fuzzy front end; consequently, it is best to acknowledge it and set about exiting this phase by obtaining the missing information
Gantt Chart	• A horizontal bar chart used in project scheduling that shows the start and end dates of tasks within a project. • A Gantt Chart is a very common way to depict project schedules due to its simplicity and visual impact.
Gate	• The decision point, generally a meeting, at which a management decision is made to allow the product development project to proceed to the next gate. • A gate has only three outcomes: "YES", "NO" or "Delay." • Gates are generally related to the timeline of the project; consequently a gate has a chronological impact. • The metaphor of a gate is part of the Gate Phase approach to project management, which has been part of product development process for many years. This approach is now more likely replace by "Agile" and other forms of iterative product development.
Gate Phase	• A product development process or methodology based on specific stages, where at the end of one stage and prior to the start of the subsequent stage there is a review to ensure that a set of criteria has been met.
Gatekeeper	• The person who or organization that holds approval. • In product development, the person who or organization that decides if the product development team can proceed with a new stage if the criteria from the previous stage has been met.
Go/No Go Gate	• A gate in the development process that allows only two outcomes: proceed or stop. • Critical in any development process, as it determines if the development meets the criteria to proceed to the next gate in the process. • A "no go" decision does not necessarily mean end of the development project.
Goal	• A purpose toward which a project is aimed. • A general direction of achievement that is not measurable. Rather, it is statement of intent; for example, "increase sales" or "accelerate product development." • We distinguish between "goal" (immeasurable achievement) and "objective" (measurable achievement). Some choose to reverse these terms.
Good Practices	• Obtained by establishing benchmarks to identify what meets a criteria of excellence. • See also "Best Practices."

Glossary

Term	Definition & Notes
High-Performance Team	• A small number of people with complementary skills who are committed to a common purpose, performance goals and approach for which they hold themselves mutually accountable — and are deeply committed to one another's personal growth and success. — from The Wisdom of Teams, by Katzenback.
Hurdle Rate	• The minimum return on investment or internal rate of return a new product must meet or exceed as it is considered for approval and proceeds through development. • A Hurdle rate can be applied to any investment.
Ideation	• Similar to brainstorming but focused on generating specific ideas about improving the performance of a product or service. • The process of idea generation generally occurs at the early stages of development.
Inbound Marketing	• The collection of all activities that provide information about the market, including market research, customer research, feedback from other organizations like sales, training and support, industry reports, competitive analysis and technology assessments. • The function responsible for obtaining and prioritizing customer requirements, then collaborating with engineering to develop innovative solutions. • Performing both tactical competitive analysis (product-to-product) and strategic competitive analysis (company-to-company).
Infant Mortality	• Failure of a new product at the early stages of use by the customer. • It can be a metric of the rate of early failure of a new product. • Not to be confused with DOA (Dead on Arrival), which is a different metric.
Innovation	• The introduction of new capabilities that create value for customers. In the business world, innovation is so only if the customer says it is — only if the customer sees value in your offering. • The ability to deliver new value to customers across the entire set of customer experiences — from sales, to delivery, to support, operations and beyond. Every "touch point" with customers provides ample opportunity for innovation and thus value creation. • It's one thing to think up an innovation; it's another to develop it and deliver it to the customer. • Innovation is clearly not limited to technology!
Intangibles	• Attributes of an offering that can't be perceived by the senses. Goodwill, brand, image and position are examples of intangibles. • The role of intangibles in the customer decision process is often underestimated leading to products that are perceived by the customers as incomplete.
Intellectual Property (IP)	• Information, formulas, designs and other knowledge that is unique to an individual or company and can be used to grow a business or improve the competitiveness of an organization. • IP takes many forms, but the main criterion is that it can provide a competitive advantage or can be sold to someone who values it.

Glossary

Term	Definition & Notes
Issue	• A known obstacle, problem or anything that has 100% probability of occurring. And when it does occur, it will cause serious or fatal disruption to a program or project.
Kaizen	• A Japanese term that represents the philosophy of continuous improvement. • More than a set of tools or processes, Kaizen is a state or mind, a commitment to continually improve and striving for perfection.
Kano, Kano Analysis	• A model developed by Dr. Noriaki Kano to characterize quality as perceived by the customer. • Generally shown as a two-dimension diagram, the model maps the degree to which the new product performs versus the degree of satisfaction of the customer. • The model assumes that all input shown was provided by the customers, and not internal personnel. • The model can be very tedious and involved.
Lagging Metrics	• See "Outcome Metrics."
Lead Customers	• Carefully selected representatives of customer groupings (segments) that have needs and requirements typical of the rest of the segment. • In some companies, Lead Customers represent customers who are facing problems ahead of the rest of the segment, providing early visibility of future customer needs.
Leading Metrics	• See "Process Metrics."
Lessons Learned	• The benefit of performing retrospects (postmortems) and implementing the improvements as a result of the process. • Lessons learned do not benefit the organization until they are applied to improve the future performance of the organization.
Life Cycle Costs	• The total costs that the customer will incur during the useful life of the new product. These costs include purchase, maintenance, repair, installations, training operations and disposal. • Generally used during the selling cycle. Customer wants to know the total cost of a new product, not just the selling price.
Market Segment	• A group of customers who display similar characteristics relevant to a business. • Market segmentation should answer three questions: – Who are the customers? – Where are they located? – What are they trying to achieve?
Market Segmentation	• The process of grouping customers with similar characteristics. • Generally the segments represent the view of the customers; that is, how they classify themselves. • A market-driven segmentation should include your products or service. It should focus on how customers group themselves. Subsequent versions of the segmentation may include your products.

Glossary

Term	Definition & Notes
Market Sensing	• Research, observations and experience develop a deep understanding of the forces acting on a market to enable the development of a vision of the future. • Sensing the market is generally the responsibility of the senior staff and involves deep participation in the affairs of a market place. • Gathering customer needs and customer requirements is not part of Market Sensing.
Market Share	• A company's sales in a product area expressed as a percent of the total sales into that market by all competitors • Market share depends on what the company defines as its market and its competitors. • If it only considers competitors as companies that offer the same product type, it will have a higher market share than if it defines its competition as including substitution products.
Market-Driven	• A commitment to serve the needs of a market. • A chosen strategy and cultural shift that enable the achievement of a set of business objectives by focusing on serving specific market segments, gaining profound insight into specific market segments, delivering superior value and earning customer loyalty.
Marketing Requirements Document (MRD)	• The process that the marketing function uses to communicate to engineering (R&D) the problems that a new product must address. This takes the form of a document called the MRD, which articulates the prioritized customer requirements. Engineering uses the MRD to generate product concepts, features and specifications • The MRD goes by various names depending on the company or industry. There is always a document that serves to communicate the prioritized customer requirements to the R&D organization.
Mean Time Between Failures (MTBF)	• The number of failures in a given period of time divided by the total uptime in the same period. • A measure of reliability that represents the average time between failures.
Mean Time to Repair (MTTR)	• The total estimated downtime of the new product for a given period divided by the number of failures or repairs performed in the same period. • Used to plan maintenance programs, calculate Total Cost of Ownership and compare reliability.
Metrics	• A standard of measurement. For example, metrics will directly measure our progress toward achieving the organization mission and group's objectives. • There are three kinds of metrics: 1. Results metrics. 2. Process metrics. 3. Customer satisfaction metrics.

Glossary

Term	Definition & Notes
Mission	• The specific task or assignment adopted by an entire group or team. The Mission describes what it does, for whom it does it and the reason it does it. • The mission statement should be specific, achievable, easy to articulate and motivating. For example: – The mission of the police is to protect and to serve. – The mission of the product team is to get a specified product out on time and meet customer requirements.
Need	• Describes what the customer says he is lacking, what he would like to do or obtain or what he wants to get. • Customer needs, as expressed by customers, are often limited to product "features" (attributes) or problems that may or may not be related to the customer end goals or to how the decision will actually be made between competing offerings. It is the reason we introduced the term "Compelling Reason to Act."
Net Present Value	• Discounting future cash outflows and inflows using today's "cost of money." • A way to normalize several investments so you can compare apples to apples. • A way to determine the viability of an investment objectively. • Used to evaluate several alternative new products and compare their financial viability.
New Product Development (NPD)	• The processes, models and tools to develop a new product. • NPD ensures speed, repeatability and concurrency. • The processes must be documented, measured and improved in order to yield long-term benefits.
New Product Introduction (NPI)	• The process of rolling out or launching a new product to the market. • The go-to-market strategy and its corresponding tactics.
Objective	• A specific, definable and measurable end-result, which is to be pursued by a team or an individual of a company within a specific period of time. • Objective and Goal are often used interchangeably. See this definition.
Offering	• A set of products, services and intangibles offered to satisfy particular customer problems. • Express what the product is from the customer's perspective. • See also "Complete Product" and "Whole Product."
Outbound Marketing	• The function that leads all activities that take the products, programs and messages to the market. • Includes managing the distribution channels, enabling the direct sales channel in all its forms, marketing communications and customer retention programs, pricing and pricing schedules, new product roll-outs, etc.
Outcome	• The measurable outputs of an action project or program. The measurement of the actual results. • It is important not to confuse "completion" of a task with "results"; they are not necessarily the same.

Glossary

Term	Definition & Notes
Outcome Metrics	• Also called "Lagging Metrics," they track the final outcome or result of a project or process.
Pipeline	• See "Roadmap."
Plan	• A detailed formulation of a program of action; for example, a Marketing Plan. • A written set of tasks, schedules and resources with defined end dates and an assessment of any barriers which might put the schedule at risk. • There are many outlines for a plan, but generally a plan answers seven critical questions: 1. What needs to be done? 2. Who will do it? 3. By when will it be done? 4. How much will it cost? 5. What are the threats and risks? 6. Why are we doing it? 7. How will we mange change?
Policy	• A definite course or method of action to guide and determine present and future decisions. • An envelope for action with clearly defined boundaries. • The organization may choose to make policy inviolate to provide a more robust framework for action.
Portfolio Management	• A process to prioritize development projects based on a set of documented criteria. • Can be used any place where a business decision must be made. • The intent is to maximize the value of the portfolio to the corporation and to ensure that corporate strategies are in full alignment with the flow of new products. • Other names include Pipeline Management, Product Pipeline, The Product Funnels and The Product Portfolio.
Postmortem	• A process to understand the effectiveness of a project and to use the information to improve future performance. • Postmortems can be done at any time during the life of a project; they are most commonly done at the conclusion. • Postmortems are only useful if the information is used to improve the future performance of the organizations. • Postmortems are also known as Debriefs, Retrospects and Lessons Learned.
Procedure	• A written series of steps followed in a regular manner to accomplish a specific task in reproducible manner. • Similar to a process, but generally assigned to a single individual. Some compare a task with an action item.
Process	• A documented, sequential list of value-added steps that can replicate a product or service to a customer. • Unlike a project, processes are designed to be repeatable. A project is a single event with a clear start and end.

Glossary

Term	Definition & Notes
Process Metrics	• Also called "Leading Metrics," they measure activities that lead to a result. They are important to track progress towards a goal. These metrics are helpful to improve processes.
Process Out of Control	• When the number of defects exceeds boundary conditions. • Thus, an engineering organization, as a process, is out of control when there are no metrics to detect the number of defects nor a system for remedial and corrective action.
Product	• Anything that can be offered to a market that represents value; i.e., demand. • Good or service that meets the requirements of a market segment. • Colloquially, a manufactured good or the tangible part of value. But this is a limiting connotation, as customers in a business-to-business environment buy solutions to problems, and such solutions are made up of tangibles, services, experiences and intangibles. Sometimes the term "whole product" is used when referring to the entire package of tangibles, services, experiences and intangibles offered to a market.
Product Benefit	• The positive outcomes or results that an offering promises a customer. • Anticipated benefits apply before the purchase. Actual benefits apply after the purchase and manifest themselves as the customer's experience with your offering.
Product Development Team	• All the individuals whose contribution will result in the delivery of a new product to the marketplace. • In some organizations the Product Development Team is defined more narrowly; for example, only the marketing and engineering professionals involved.
Product Funnel	• See "Pipeline."
Product Life Cycle	• The four stages that all new products go through from introduction to end-of-life: introduction, growth, maturity and decline. • Duration of time from the earliest proposal to removal from the order entry system (phase out). A series of phases products follow from earliest conception to end-of-life • Product Life Cycle has many applications in product companies; such as tracking the product portfolio, ensuring profitability (gross margin time number of units), resource allocation, etc.

Glossary

Term	Definition & Notes
Product Management	• The process and strategy to mange the life cycle of a product such that it meets profitability targets. • The role of Product Management is to define, develop, deploy and maintain products and services that: – Provide more value than the competition. – Help build a sustainable competitive advantage. – Deliver financial benefit to the business. • Six of the key disciplines of product management are: 1. Market Research 2. Product Definition and Design 3. Project Management 4. Evangelize the Product 5. Product Marketing 6. Product Life Cycle Management
Product Manager	• The person assigned the responsibility and accountability for leading all the various activities that concern a particular product along its life cycle. • Sometimes "Brand Manager", particularly in consumer products. • Generally the Product Manager has P&L (Profit and Loss) responsibility for the product line. Some companies do not give P&L responsibility, but assign a "pro-forma" P&L to signify the importance of the job. • The Product Manager is responsible and accountable for achieving the business objectives of the product line.
Product Portfolio	• See "Pipeline."
Product Requirements Document (PRD)	• See "Marketing Requirements Document."
Program Evaluation Review Technique (PERT)	• A form of network analysis used to estimate the duration of a project.
Program Management	• The business function that assures the clean execution of corporate strategies, where a program is a group of related projects managed in a coordinated way to obtain the benefits and control not available from managing them individually. • The investment in program management is made to: – Reduce the variability of return from an investment. – Ensure alignment between strategy and execution in order to meet the business objectives of the enterprise. – Deal with change in a dynamic environment to ensure business results. • Program Management is not the same as "Project Management."
Program Manager	• A person with the skills, knowledge and abilities to lead a cross-functional team in the development of a product with a successful outcome such as profitable time-to-market. • This definition applies to the context of product development — your organization may use this title differently within a different context.

Glossary

Term	Definition & Notes
Programmatic Customer Visits	• A structured set of customer interviews where the participants share a common set of topics to be discussed. Further, the participants agree to a regular schedule of trips to the field to conduct the interviews. • A common way to capture, prioritize and archive the information from the customers.
Qualify Function Deployment (QFD)	• Quality Function Deployment is a structured methodology that uses a set of matrices for linking market requirements to the entire product development effort. • QFD will not define a product for you; it is simply a tool to manage information about the new product so that the integrity of the customer requirements is not jeopardized. • QFD is also called the House of Quality due to the appearance of the matrices. • QFD originated in Japan as a way to manage all the features and their interactions of a new product from start to finish.
Quality	• Perception of the degree to which a product or service meets customers' expectations. • Quality is whatever your customer says it is. • Quality is an attribute of the product that generates customer satisfaction. But it is also a perceived attribute, and therefore is somewhat subjective. • Quality Standards are approved, documented rules that define one or several dimensions of quality for a particular application or set of customers.
Rapid Prototyping	• Any of a variety of processes that avoid tooling time in producing prototypes. Generally these are nonfunctioning. • Rapid Prototyping exists to test the feasibility of a particular design with customers to reduce time-to-market.
Remedial Action	• Activities to provide an immediate resolution in order to solve a problem to the satisfaction of the customer. • Without regard to permanent solutions.
Request for Proposal (RFP)	• A process used by larger companies, government entities, military and universities to invite suppliers to bid on a specific acquisition of products, services, maintenance, etc. as a package. The submitted RFPs are then used by the buying entity to evaluate the merits and make an informed decision to purchase.
Request for Quotation (RFQ)	• See "Request for Proposal (RFP)."
Requirement (Customer)	• A brief and clear statement expressing what the customer will be able to experience as a result of owning and/or using the offering. • Product and customer requirements are two ways to express the same thing. • Requirements must be short, unambiguous, and able to be verified
Requirement (Product)	• A brief and clear statement expressing what an offering will have such that the needs of the customer will be met. • Product and customer requirements are two ways to express the same thing. • Requirements must be brief, unambiguous, and able to be verified.

Glossary

Term	Definition & Notes
Results	• The measurable achievement of an endeavor; for example, the results of a winning product is the achievement of the profitability target. • The outcome of an endeavor.
Rev (Review or Revise)	• To review or revise, used in software development to track the number of revisions that have been made to a specific application. • Revision can also be applied to documents to track version control.
Reverse Engineering	• The process of copying a product by analyzing every part of a competitor's product and then incorporating its features into one's own products. • Generally not a good thing to do.
Risk	• Something that can harm your program and should not happen. • The probability that something may go wrong with your project. • There are many types of risks that can interfere with your product development.
Risk Management	• The process to identify, prioritize, plan and manage risks in a project. • Risk Management starts at the earliest stages of a project and continues through completion.
Risk Management Plan	• The documented set of prioritized risks and corresponding mitigation plan. • The Risk Management Plan is constantly monitored.
Roadmap	• The manifestation of the product strategy for a product line or a company. • The outline of the future products versus the time it takes the organization to develop to achieve its business objectives. • Organizations may use terms like "Funnel", "Product Funnel", "Product Roadmap", "Pipeline" or "Product Portfolio." We suggest all of these terms are synonymous.
Simultaneous Engineering	• See "Concurrent Engineering"
Situation Analysis	• The current snapshot of the factors confronting a team, company or organization in a complex situation. • A short list of items, observations or data that are highly relevant to the topic at hand and are presently important to consider. • Examples include the Situation Analysis of a market and the Situation Analysis in a Product Proposal.
Sponsor, Team Sponsor. Team Mentor	• An Executive of the company who agrees to support a team in order to ensure its success. • The sponsor must have the time, willingness and skills to assist the team in a measurable way.
Strategic Alignment	• The endeavor to align all stakeholders on a single strategic direction such that their day-to-day activities fully support the endpoint. • Strategic Alignment complements (but is different from) "Strategy Formulation" or "Strategizing", which is the process of synthesizing and articulating a particular strategy.

Glossary

Term	Definition & Notes
Strategic Formulation (Strategizing)	• The act of formulating a strategy, different from the process of strategic planning. • Strategizing is the thought process or brainstorming needed to synthesize a strategy given a clear understanding of the market.
Strategic Initiative	• A program that, when completed, will deliver visible results in clear support of an overall strategy. It's defined to clearly move the organization towards the achievement of the corporate or group strategic goal. • Strategic Initiatives possess the following attributes: – Deemed critical to the success of the organization — non-negotiable and successful implementation is the only outcome. – In general, assigned to a member of the executive team as part of his MBO. – In general, implementation is carried out by one or more small teams that are given a clear charter, including deliverables, metrics, desired outcomes, time lines, roles and responsibilities. The executive is still accountable for measurable results. – Periodic reviews are included to ensure progress and accountability. – Fulfilled through programs and projects. – Span more than one fiscal year or are part of the yearly operational plan.
Strategy	• The approach that a team or company chooses to achieve a set of business objectives. • The approach guides the choices, behavior and direction of an organization.
Strategy (Corporate)	• A set of Strategic Initiatives made up of programs and projects that, when implemented, will fulfill the vision of the company. • This is the means by which you connect everyone's day-to-day activities to the vision of the company.
Strategy (Year Around Planning)	• A model of strategic planning that, unlike the yearly cycle of "business planning", assumes that strategy is a constant endeavor. • Usually characterized by a rigorous review and update of the strategy twice or three times per year.
Superior Value	• Better than your competition. Always defined by your customers, based on their comparison of you with your competitors. • Superior Value shows when a customer selects your products or solutions over that of your competition.
Switch-Over-Cost	• A term used to describe the mental process to shift one's cognitive resources from one topic to a completely different one. • Research done by Pattison, Kermit and Worker in "The Cost of Task Switching" shows that it may take up to 23 minutes to return to the same level of concentration after an interruption.
Team Charter	• See "Charter."
Team Sponsor, Team Mentor	• See "Sponsor."

Glossary

Term	Definition & Notes
Theory of Inventive Problem Solving (TRIZ)	• A Russian acronym that refers to a methodology for solving a difficult problem whereby multiple alternative solutions are defined. • The method is supposed to improve creativity by enabling the team to think of many solutions to a single problem. Generally a multi-discipline methodology to tap of the creativity of different individuals.
Timeline	• The schedule or the amount of time plus the activities needed to deliver a set of results. • A Gantt chart showing the chronological progression to the completion of a project.
Time-to-Market	• A measure of the speed by which a new product is developed. • Measured from the official start of the development until the product is ready to be purchased by a customer. • A critical measure to ensure competitiveness and success in the market place. • Various companies use different start and end points to calculate their time-to-market.
Time-To-Results (Time-To-Profit or Time-To-Cash)	• That point in the product life cycle where a target result that determines the success of the product has been met. • The success of a product may be measured by margin, revenue, market share or other strategic goals. • The time-to-results point is determined no later than the creation of the product proposal.
Total Cost of Ownership(TCO)	• The aggregate of all the costs during the life of a product, from acquisition to disposal.
Use Case	• Describes a specific activity or set of activities performed by a customer with a device or product. • Use cases are created with specific goals to understand the interaction between an operator and the equipment.
Value	• The customer's perception of what he wants to happen in a specific use and situation as the result of using a company's offering. • There is value in using the product or service — i.e., deriving a benefit as a result of use. There is also value in owning the product or service — i.e., deriving an emotional benefit just from owning the product or service.
Value Proposition	• A statement describing what resulting experience the customer will have, articulated in a way that communicates clearly and briefly the benefits, costs and competitive advantage of a particular offering. • The value proposition must be compelling and differentiated as perceived by the customer. • The resulting experience includes the outcomes or results that the customer will experience. • Whole products deliver value propositions.

Glossary

Term	Definition & Notes
Values	• A set of deeply held beliefs that enable employees to act without the need to seek management approval. • Values are never intentionally compromised. • A common set of social principles that guide the behavior and decisions of each person and the organization.
Vision	• The vivid description of the end destination and results of a group effort. • The distant lighthouse that provides direction to the completion of the team's journey. • A vision statement evokes concrete images. It is compelling and prompts forward progress. It is succinct and memorable.
Voice of the Customer (VOC)	• A process that uses structured, in-depth interviews for eliciting customer requirements. • VOC is also used as a generic way to talk about customer needs or customer requirements. • VOC by itself does not yield customer needs; other processes have to be used in conjunction to process the input from the customer, turn it into product features and eventually into a new product.
Voices	• Verbatim quotes from customers that provide insights as to their needs. • These voices are processed to extract requirements.
Whole Product	• The set of product, services, ancillary parts and intangibles needed to ensure that the customer will achieve his compelling reason to act (CRTA). • The actual product is a subset of the whole product. The whole product is what the customers buy. • Ignoring dimensions of the whole product explains why so many companies fail to penetrate new markets. • See also "Complete Product" and "Offering."

FMEA Process Map from SEMATECH, Inc.

Overview

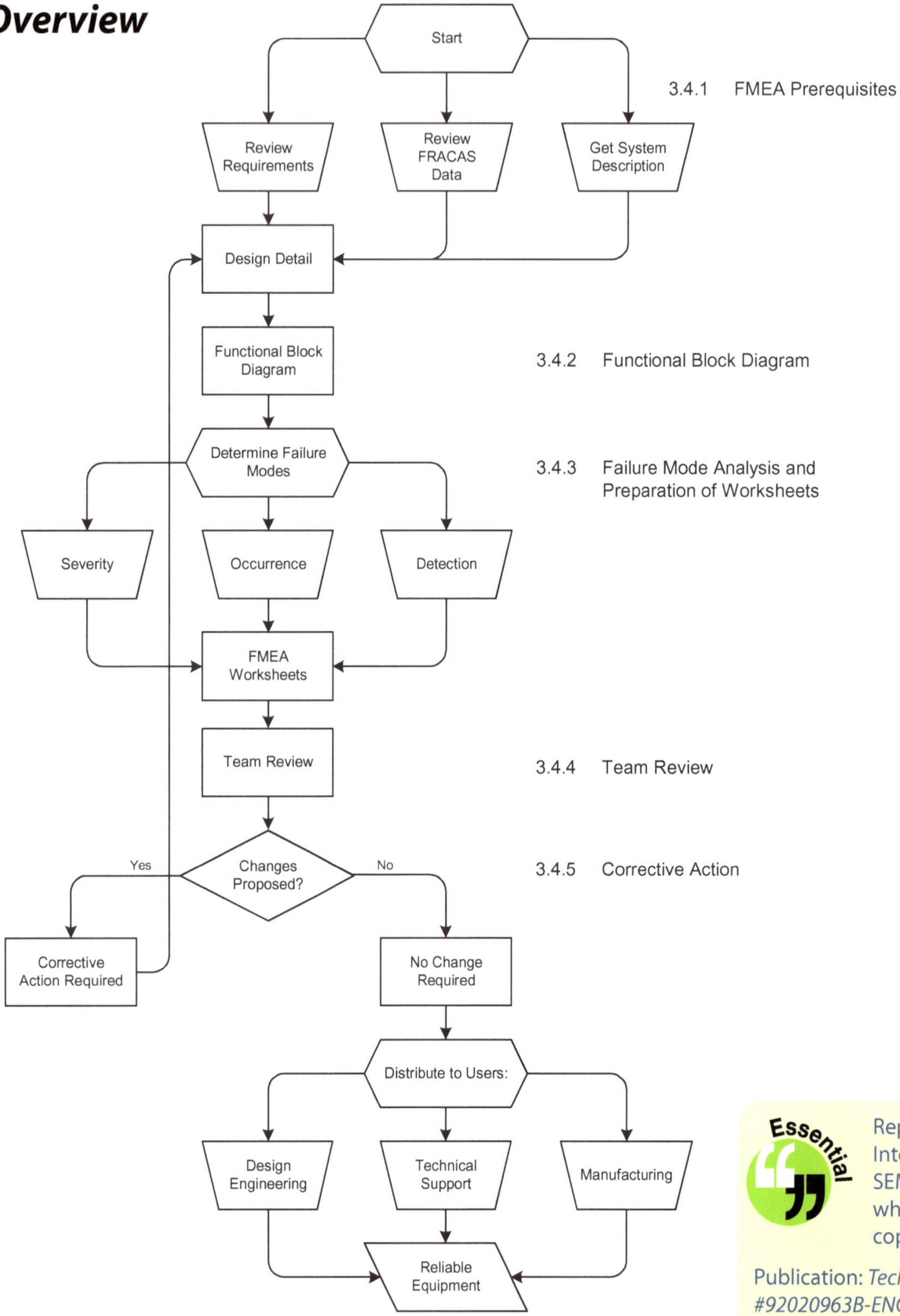

Figure 3 FMEA Process

Technology Transfer #92020963B-ENG

Templates

Instructions

These templates are for you to use and share. Throughout this guidebook, you have seen real world tools and examples. Any page with a template icon indicates there is a correlating fillable form in this section. For more resources and future updates, visit us at: www.rapidinnovation.com.

Templates with Examples

More Tips for Users of the Interactive PDF Version

You can view the complete list of templates with the Show/Hide Templates bookmark on the left Navigation Panel. Double click any template you want to work on. You can also open a template by double clicking the white pushpin icon on the specially marked example pages.

Once opened, these templates are enabled for use in Adobe Reader or Adobe Acrobat — you can fill them in with your own information and share with peers and colleagues. When you open original templates from your guidebook, they will be fresh. But you can save versions for future access — whether you've completed the template, partially filled in your information or left it blank. Saved versions remain enabled for revisions and collaboration. When you close the template, you will return to this guidebook.

These enabled templates also provide you with Comment & Mark-up Tools. Just as you might with a hard copy, you and your team can highlight text, add notes and make custom annotations.

Communication Matrix

Use the communication matrix template to identify those individuals who must be kept informed about your project. The template shows various functions and positions within a typical project.

KEY DECISION MAKER: The one person who will make all the important decisions about your project. Generally a manager or executive in your organizations. In most cases, the person who is providing the funding for your project. You must create a very clear channel of communication with this person.

PROJECT SPONSOR OR MENTOR: A manager or executive who commits to helping your team and ensuring the success of your project. The sponsor or mentor will facilitate the smooth implementation of your project; for example remove obstacles, obtain additional resources and provide advice and counsel. In some organizations "sponsor" may mean the same thing as the key decision maker.

PROJECT CUSTOMERS: Those individuals or organizations that will derive the value your project generates. They could be internal or external customers, paying or otherwise. Do not confuse "stakeholder" with "project customer."

STAKEHOLDER: Someone who has a vested interest in the success of your project; for example your manager or a board member.

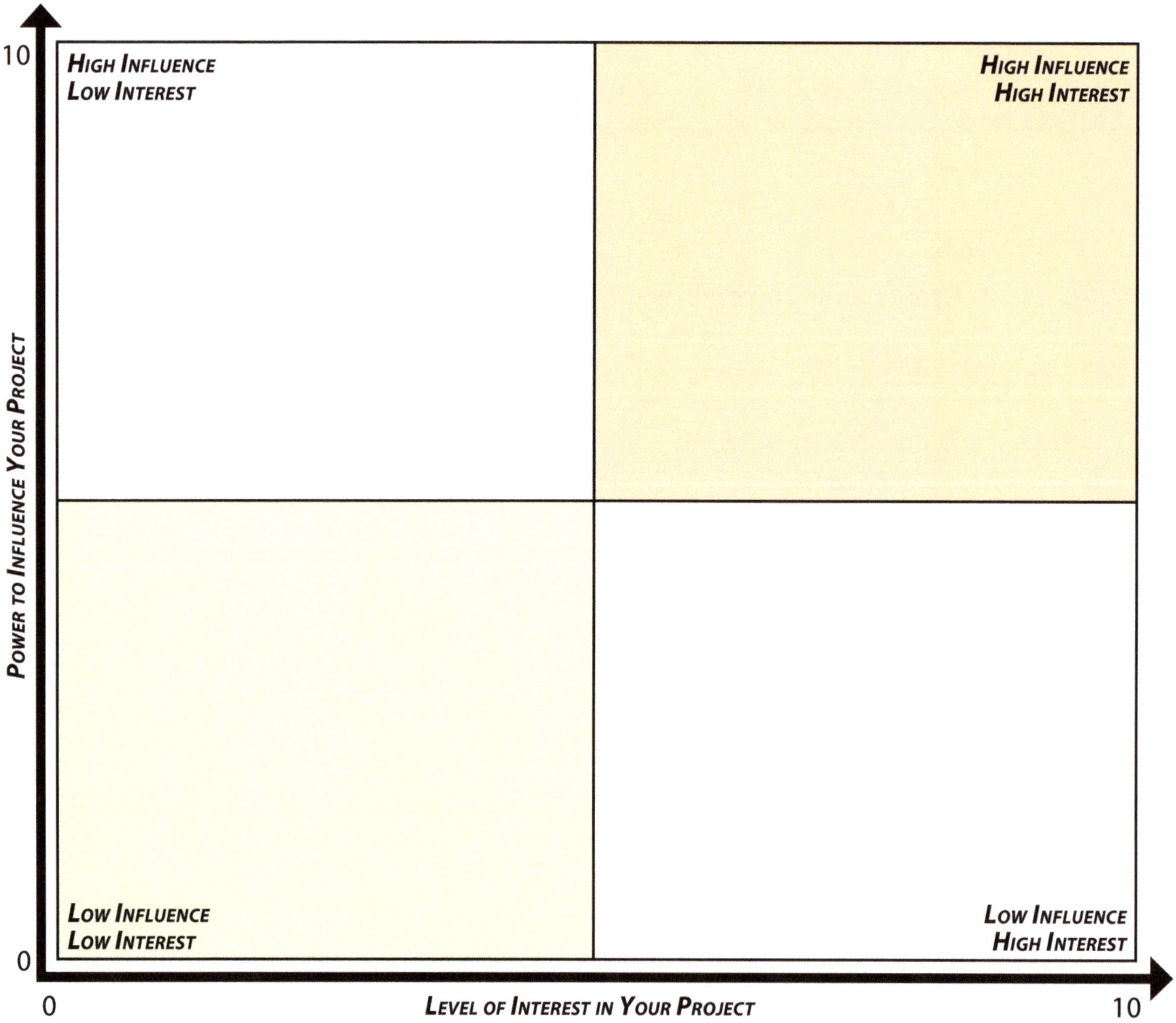

Flexible Project Management for Product Development

Do it now. Do it right. Deliver results. Templates for the Real World.

Communication Decision-Making Tree

Project Name: ______________________ Date: ______________

Project Leader: ______________________

Project Sponsor or Mentor

Project Team Members

Name: ______________
Title/Function: ______________
Contact Info: ______________

Name: ______________
Title/Function: ______________
Contact Info: ______________

Name: ______________
Title/Function: ______________
Contact Info: ______________

Name: ______________
Title/Function: ______________
Contact Info: ______________

Name: ______________
Title/Function: ______________
Contact Info: ______________

Name: ______________
Title/Function: ______________
Contact Info: ______________

Name: ______________
Title/Function: ______________
Contact Info: ______________

Name: ______________
Title/Function: ______________
Contact Info: ______________

Name: ______________
Title/Function: ______________
Contact Info: ______________

Name: ______________
Title/Function: ______________
Contact Info: ______________

Name: ______________
Title/Function: ______________
Contact Info: ______________

Name: ______________
Title/Function: ______________
Contact Info: ______________

Project Customers

- Those who will derive value or from the project.
- Those who will purchase the end result or your project.
- Those who benefit from your project.
- Those who will use what your program delivers.

Key Decision Maker

Other Key Stakeholders

Outside Suppliers (Outsource)

Code	Responsibility
A	Approve
R	Review

Code	Responsibility
C	Create
D	Deliver
I	Inform

Code	Responsibility
IN	Provide Input
N	Notified
L	Leads a Team or Sub-Team

Charter Tool

1. Project name:
    - ❖
    - ❖
    - ❖
    - ❖
    - ❖
2. Short description of your project:
    - ❖
    - ❖
    - ❖
    - ❖
    - ❖
3. What is the alignment of this project to the business strategy of your company?
    - ❖
    - ❖
    - ❖
    - ❖
    - ❖
4. What are the project deliverables?
    - ❖
    - ❖
    - ❖
    - ❖
    - ❖
5. How will you know when the project is successful?
    - ❖
    - ❖
    - ❖
    - ❖
    - ❖
6. Key assumptions about this project.
    - ❖
    - ❖
    - ❖
    - ❖
    - ❖

7. Key constraints; i.e., what are the boundary conditions?

- ❖
- ❖
- ❖
- ❖
- ❖

Roles and Authority

8. Who is the Project Manager?

- ❖
- ❖
- ❖
- ❖
- ❖

9. Who is the Sponsor?

- ❖
- ❖
- ❖
- ❖
- ❖

10. Who will make the key decisions?

- ❖
- ❖
- ❖
- ❖
- ❖

11. Other stakeholders?

- ❖
- ❖
- ❖
- ❖
- ❖

Approvals

Management Sponsor: ____

Project Manager: ____

Funder ($): ____

Kick-Off Meetings Planning Tool

Name of the proposed Kick-Off Meeting: ____________________

1. What is the purpose of the Kick-Off Meeting?

- ❖
- ❖
- ❖
- ❖
- ❖
- ❖
- ❖

2. Purpose of this set of visits:

- ❖
- ❖
- ❖
- ❖
- ❖
- ❖
- ❖

3. What are the tangible results that you must have from this meeting — i.e., the outcomes?

- ❖
- ❖
- ❖
- ❖
- ❖
- ❖
- ❖

4. What are the methods you will use to achieve the objectives of the Kick-Off Meeting?

- ❖
- ❖
- ❖
- ❖
- ❖
- ❖
- ❖

Contact the author: flexpm@rapidinnovation.com.

5. What are the roles for the meeting?

- ❖
- ❖
- ❖
- ❖
- ❖
- ❖
- ❖

6. What will happen the day after the meeting?

- ❖
- ❖
- ❖
- ❖
- ❖
- ❖
- ❖

7. What will happen a week or two after the meeting?

- ❖
- ❖
- ❖
- ❖
- ❖
- ❖
- ❖

8. How will we evaluate the success of the meeting?

- ❖
- ❖
- ❖
- ❖
- ❖
- ❖
- ❖

9. Who must attend the meeting?

- ❖
- ❖
- ❖
- ❖
- ❖
- ❖
- ❖

Contact the author: flexpm@rapidinnovation.com.

Flexible Project Management for Product Development

Do it now. Do it right. Deliver results. Templates for the Real World.

Gantt Chart and WBS

	Level From Your WBS			Months						
#	L1	L2	L3							
1										
2										
3										
4										
5										
6										
7										
8										
9										

Responsibility Matrix

Project Name: ______________________ Date: ______________

______________________ Project Leader: ______________

	Project Sponsor	Functional Manager	Project Manager	Project Team	Steering Committee
Project Definition					
Business Requirements					
Budgeting					
Status Reports					
Risk Management					

Code	Responsibility
A	Approve
R	Review

Code	Responsibility
C	Create
D	Deliver
I	Inform

Code	Responsibility
IN	Provide Input
N	Notified
L	Leads a Team or Sub-Team

RACI

Function or Task: ______________________________

No.	Task	R Review	A Approve	C Contribute	I Inform

Analogous Estimate

Cost Estimate

Company Name: ____________________ Project Number: ____________________ Estimate Number: __________

____________________ Project Manager: ____________________ Date: __________

Line Item	Type of Labor						Total Cost of Labor	Overhead	Billable Cost
	Engineering Rate: $100		Technician Rate: $50		Helper Rate: $30				
	Hours	@ Rate	Hours	@ Rate	Hours	@ Rate			

Develop the Cash Flow or Time-Phased Budget

Cash Flow

Company Name: ____________________ Project Number: ____________________ Estimate Number: __________

______________________________ Project Manager: ______________________________ Date: __________

Line Item	Jan	Feb	Mar	Apr	May	Jun	Jul	Aug	Sep	Oct	Nov	Dec	Ttl

Focussed Risk Management

Types of FMEA	Program Level	Project Level
Financial (Business Case)		
Marketing (The Market Place)		
Engineering Design (Software and Hardware)		
Manufacturing (Process)		
Supply Chain (Suppliers)		
Service and Repair (Field service, repair and maintenance)		

Types of Risk Management to be Performed

	Level of the Risk Management Plan				
Types of FMEA	**Program Level**	**System Level**	**Subsystem Level (Module)**	**Component Level**	**Pre-Phaseout**
Financial (Business Case)					
Marketing (The Market Place)					
Engineering Design (Software and Hardware)					
Manufacturing (Process)					
Supply Chain (Suppliers)					
Service and Repair (Field service, repair and maintenance)					

PI Matrixes

Contact the author: flexpm@rapidinnovation.com.

Tabular PI Matrix

Item	Probability	Impact	PI Factor	General Notes

Contact the author: flexpm@rapidinnovation.com.

Documenting Your Risk Management Plan

Template 1

Risk Management Plan										
Work Package or Task	Risk Event No. and Description	Probability	Impact	Plan			Actual			Owner
				Risk Event Status (P x I)	Criticality	Impact on Other Risks	Preventive	Trigger Point	Contingent	

Contact the author: flexpm@rapidinnovation.com.

Documenting Your Risk Management Plan

Template 2

Risk Management Plan

Work Package or Task	Risk Event No. and Description	Owner	Criticality	Plan			Actual			
				Preventive	Trigger Point	Contingent	Status	Date of Impact	Risk Impact	Recovery Action

Documenting Your Risk Management Plan

Template 3

Risk Management Plan											
				Plan			Actual				
Cause of Risk	Description of Risk	Description of Impact	Probability that Risk Will Happen	Impact of Risk if it Does Happen	Ranking Based on P and I	Trigger Point	Contingency Plan A	Contingency Plan B	Risk Owner	Comments	

Examples: Documenting Your Risk Management Plan

Example for Template 1

Work Package or Task	Risk Event No. and Description	Probability	Impact	Plan			Actual			Owner
				Risk Event Status (P x I)	Criticality	Impact on Other Risks	Preventive	Trigger Point	Contingent	
Develop blueprints	Draftperson not available due to illness	2	7	14	Critical	Yes Engineering	Hire another draftperson	Draftperson calls in sick	Borrow draftperson from another department	Fred Smith
Stone work completed	Stone mason from Europe unable to arrive on time	5	6	30	Critical	No	Coordinate travel arrangements well in advance	The stone mason is not on site 6 days prior to start	Hire local stone mason to start positioning the material	May Jones

Example for Template 2

Work Package or Task	Risk Event No. and Description	Owner	Criticality	Plan			Actual			
				Preventive	Trigger Point	Contingent	Status	Date of Impact	Risk Impact	Recovery Action
Develop Blueprints	Draftperson not available due to illness	Marsha L.	C (Critical)	Hire another draftperson	Draftperson calls in sick	Borrow draftperson from another department	Draftperson has announced her resignation	September 2, 2005	Two-week delay	Activate preventive (hire)/Stage contingent (borrow)

Example for Template 3

Cause of Risk	Description of Risk	Description of Impact	Probability the Risk Will Happen	Plan			Actual			
				Impact of Risk if it Does Happen	Ranking Based on P and I	Trigger Point	Contingency Plan A	Contingency Plan B	Risk Owner	Comments
Draft-person is not available due to illness	Causes a delay in the creation of the schematics	Delays entire project by the same number of days	60%	$50,000 loss	#5	Draft-person calls in sick	Borrow a draft-person from another division	Hire a temporary draft-person	Fred A.	Will need to train either replacement

Monitor Your Risk Plan Using the Critical Questions of Proactivity

This list can also be used as the agenda for your Risk Management Review meeting.

What has changed since we last met?

- ❑ What important changes have taken place that impact our project?
- ❑ Are there any new risks that must be added to the Risk Management Plan?
- ❑ Who are the risk owners for the newly identified risks?
- ❑ Are there any old risks that are obsolete and must be deleted from the Risk Management Plan
- ❑ What changes have happened to the current list of risks?
- ❑ What other changes have take place?

What are the reasons for the change or variations?

- ❑ What has caused the new risks?
- ❑ What caused the changes to existing risks?
- ❑ What cause the deletion of risks?
- ❑ What are the implications of the changes?
- ❑ Are there any trends that will impact our Risk Management Plan?
- ❑ What will happen if the new trends continue?
- ❑ What should we do right away to mitigate the trends?

What are the specific changes to our current Risk Management Plan?

- ❑ Add new risks?
- ❑ Delete risks?
- ❑ Changes to current risks?
- ❑ Changes in responsibilities?

Status Report (Progress Report)

Project Name: ______________________ Date: ______________

Project Leader: ______________________________________

Item	Gantt Chart Schedule					
Visit customer site.						
Prepare the plan.						
Develop costing.						
Obtain initial permits.						

Current Issues: ______________________ Response: ______________________

Risk Management: ______________________ Response: ______________________

Action Required: ______________________ Response: ______________________

Project Change Request

Project Name: ______________________________ Date: ____________________

Project Leader: ____

Detail each change requested and its impact on the project scope and quality:

1. ____

____

____

2. ____

____

____

3. ____

____

____

4. ____

____

____

Name of person originating the request: ______________________________

Describe any emergency action needed: ______________________________

Type of change: ❑ Minor — requires no planning and minimal impact on the budget

❑ Major — requires planning and possible high impact on the budget

Is the change customer funded: ❑ Yes ❑ No

Any additional relevant information: ______________________________

____

____

____

____

Approvals ____

____

____

Selecting the Right Project to Debrief

Use this simple tool to select the right project. Develop the template to track, prioritize and optimize the results from debriefs.

Project Candidate	Business Case	Focus (Topic)	Plan Leader	Target Date	Deployment Leader
Selection of projects is based on a clear criteria, which is applied to all candidates.	The business rationale that justifies the expense in resources to perform the debrief.	The specific area of focus for the debrief. The desired outcome will dictate the focus.	The name of the person responsible for leading the planning and execution of the debrief.	The intended date when the main session of the debrief is planned.	The name of the person who is responsible to lead the implementation of the results of the debrief.

Contact the author: flexpm@rapidinnovation.com.

A Well-Documented Action Plan from a Debrief

Item	Background Info	Action Item	Responsibility	Timeline

Contact the author: flexpm@rapidinnovation.com.

Debrief Planning Tool

To be filled out by the person responsible for organizing the upcoming debrief.

Item	Description

Debrief Planning Tool by Focus (Topic)

This is a comprehensive form to enable conversations. It may be too complex for some projects or programs, so we suggest that you customize this form. Do not send this form to team members! This form is to help you facilitate the meeting, but the real value is the dialog.

1. Planning

2. Resources

3. Scheduling

4. Performance of Project or Program Management

5. R&D / Technology Management

6. Communications

Contact the author: flexpm@rapidinnovation.com.

7. Managers and Executives

8. Product

9. Marketing

10. Manufacturing (In Software Development, use Documentation or QA)

11. Virtual or Distributed

12. Service and Repair

13. General Questions

Planning Checklist for Deploying a New Tool

Deploying a tool is indeed a project. Consequently, ensure that you have a deployment plan, a schedule, budget, etc. Working with your Core Team, or with the peers who have agreed to help, you can develop a tactical plan that will ensure a successful deployment.

The list below is intended to help you formulate your own plan. You may need to add your own items to the list. Also note that the list below is in no particular sequence. The sequence will be determined by your specific needs and schedule. Finally, the list is not rigid. Use you judgment and do what makes sense; you know your organization better than most.

Planning Checklist

- ❑ Schedule for the deployment defined, documented and approved.
- ❑ Training materials for employees developed and documented.
- ❑ Training materials produced (hard & soft copies), available for use.
- ❑ Training materials placed in the intranet or Web page.
- ❑ Trainers on the new tool or method identified and informed of their duties.
- ❑ Training of employees scheduled and employees notified.
- ❑ Training of managers and supervisors scheduled.
- ❑ Feedback mechanism in place.
- ❑ Metrics to measure the progress and eventual success of the deployment identified and ready to use.
- ❑ Coaching and hand-holding for managers, employees and stakeholders defined and ready — including those who will do the coaching.
- ❑ Support (assistance) available; for example, facilitators, coaches, tech support, etc.
- ❑ Obstacle removal in place.
- ❑ Templates, checklists and other tools ready and available.
- ❑ Information on our Web page done.
- ❑ Processes that will be impacted by the deployment identified and corrected.
- ❑ Changes to our quality manual completed.
- ❑ Manuals, guides and other reference materials that will be needed by the users ready and available.
- ❑ Executives briefed and ready to fully support the deployment.
- ❑ Human resources (HR) informed.
- ❑ Training department informed.

Planning Document for Deploying a New Tool

This tool may help you secure assistance from your executives. It is also a good point to start the process of deployment. It may even become a standard template for deployment of future tools and templates. You will want to modify it to meet your specific objectives and adapt to the culture of your organization.

1. Tool or methodology to be deployed

2. Objective of the tool deployment

3. Who is the "expert" on this tool?

4. What documentation is available?

5. Are there any external resources available?

6. What are the risks involved in the deployment?

7. How will we "pilot" the tool?

8. What will be our communication plan?

9. What will we measure to assure us that the deployment has been successful?

Contact the author: flexpm@rapidinnovation.com.

FREE BONUS

How to Get Your Interactive Version of this Guidebook

Register Your Book

Thank you for purchasing *Flexible Project Management for Product Development*. Many find the interactive PDF version of this guidebook to be a valuable addition because the cross-reference links are live and all the templates can be filled out electronically and used to collaborate with your team members around the world. Once you enter the information, every template becomes a separate PDF file for sharing and collaboration. Furthermore, the electronic version of this guidebook can also be stored in any portable device that accepts PDF files such as tablets, laptops and even smartphones.

To obtain your free interactive PDF copy of *Flexible Project Management for Product Development*, follow these simple steps:

1. Send an email to flexpm@rapidinnovation.com with "**Registering FlexPM book**" in the subject line. In the body of your email, include your name plus the following registration code: **FXP4612E**.
2. In a few days, you will receive an email notice that your book has been registered with either an attached copy of your free interactive PDF or a link to download your copy.

If you have any questions during registration, please email the publisher at: flexpm@rapidinnovation.com.

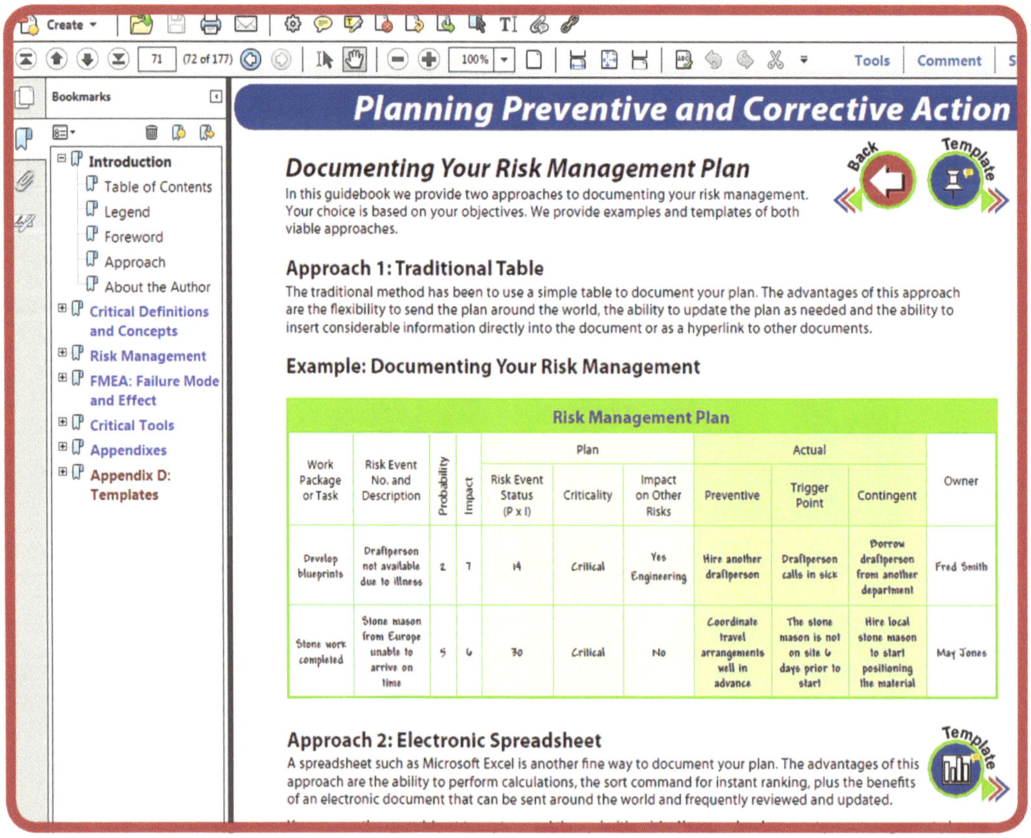

Bookmarks

Introduction
Table of Contents
Legend
Foreword
Approach
About the Author
Critical Definitions and Concepts
Risk Management
FMEA: Failure Mode and Effect
Critical Tools
Appendixes
Appendix D: Templates

Planning Preventive and Corrective Action

Documenting Your Risk Management Plan

In this guidebook we provide two approaches to documenting your risk management. Your choice is based on your objectives. We provide examples and templates of both viable approaches.

Approach 1: Traditional Table

The traditional method has been to use a simple table to document your plan. The advantages of this approach are the flexibility to send the plan around the world, the ability to update the plan as needed and the ability to insert considerable information directly into the document or as a hyperlink to other documents.

Example: Documenting Your Risk Management

Risk Management Plan

Work Package or Task	Risk Event No. and Description	Probability	Impact	Plan			Actual			Owner
				Risk Event Status (P x I)	Criticality	Impact on Other Risks	Preventive	Trigger Point	Contingent	
Develop blueprints	Draftperson not available due to illness	2	7	14	Critical	Yes Engineering	Hire another draftperson	Draftperson calls in sick	Borrow draftperson from another department	Fred Smith
Stone work completed	Stone mason from Europe unable to arrive on time	5	6	30	Critical	No	Coordinate travel arrangements well in advance	The stone mason is not on site 6 days prior to start	Hire local stone mason to start positioning the material	May Jones

Approach 2: Electronic Spreadsheet

A spreadsheet such as Microsoft Excel is another fine way to document your plan. The advantages of this approach are the ability to perform calculations, the sort command for instant ranking, plus the benefits of an electronic document that can be sent around the world and frequently reviewed and updated.

Other Titles You May Be Interested in Owning

THE SERIES

Risk Management and FMEA for Product Development

"If you are not managing risk, you are not managing your project." This is most applicable when it comes to managing risk in product development. Technology changes, customer requirements vary, schedules are always in flux, and your suppliers are in turmoil. What if you could predict what could go wrong? Better yet, what if you could develop a plan to eliminate or mitigate your risks?

Risk Management is the ability to anticipate risks and develop plans to manage them; thus, creating more predictable schedules, improving reliability and customer satisfaction. In many cases, risk management prevents what otherwise might be a catastrophic failure of your development program.

This book was written by product developers with many years of experience in the real world, where traditional text book approaches fail. This how-to guide understands the pressures placed on you, the need to get to the point quickly and to show you step-by-step how to apply risk management on product development projects.

The Voice of the Customer for Product Development

Your task is to create differentiated value for your customer in order to achieve the right gross margin at minimum COGS. This guidebook is based on experiences with many technology development teams around the world — all of them focused on achieving the right volume, margin and profitability. You will be introduced to a customer-driven methodology that has been perfected over the last 30 years.

The mission of this guidebook is to create awareness of the framework and motivation needed to successfully understand customer needs, process the information and put it to use to develop differentiated products that achieve the right level of profit. This guidebook methodically walks you through the tools, templates and methods for obtaining, processing and using customer input — like no other of its kind, it is based on real projects and it is documented to provide maximum guidance to enable application.

You will be able to increase the value of your offering offerings and create defensible market segments. You will learn how to identify opportunities by interviewing the right customers. You will also learn how to transform these opportunities into requirements and then into product concepts that cast an optimum balance between usefulness, usability and desirability. The overall objective is to create the capacity for improved productivity!

Debriefs and Postmortems for Product Development

Studies show that debriefs, or postmortems, are the best way to improve the productivity of teams — better than training, books, seminars and other tools. Your ability to methodically identify areas of improvement from development programs and immediately apply them to upcoming efforts will result in improved productivity, more enthusiastic teams and improved morale in your development organization.

This step-by-step guidebook will help you avoid the common mistakes many organizations make that block them from obtaining the full benefits from debriefs. You will learn how to prioritize improvement changes and how to deploy them through the organization in a way that it is embraced by all.

www.ingramcontent.com/pod-product-compliance
Lightning Source LLC
LaVergne TN
LVHW070947210926
848482LV00002B/121
9780981759531